Backyard
FRUITS
& BERRIES

Backyard
FRUITS
&BERRIES

How to grow them better than ever

by
Diane E. Bilderback and
Dorothy Hinshaw Patent

Rodale Press, Emmaus, Pennsylvania

Printed in the United States of America on recycled paper, containing a high percentage of de-inked fiber.

Book design by Anita G. Patterson
Illustrations by David E. Bilderback and Dorothy Hinshaw Patent

Library of Congress Cataloging in Publication Data
Bilderback, Diane E.
    Backyard fruits and berries.

    Includes index.
    1. Fruit-culture. 2. Organic gardening. 3. Berries.
I. Patent, Dorothy Hinshaw. II. Title.
SB357.24.B55   1984   634   84-8425
ISBN   0-87857-509-X   hardcover

2 4 6 8 10 9 7 5 3 1   hardcover

This book is dedicated to Dr. Nancy Callan and to all other patient researchers who are willing to give years of their lives to help unravel the mysteries of fruit production and who are willing to forego more glamorous and immediately rewarding careers in exchange for the satisfaction of providing improved varieties of fruits and fruit rootstocks for everyone to enjoy.

CONTENTS

ACKNOWLEDGMENTS

We wish to acknowledge the help of several generous people who gave of their time and expertise in helping us with this book. First of all, we want to thank Dr. Nancy Callan of the Western Agricultural Research Station, Montana State University, for reading and commenting on chapters 1 and 6 and for providing us with much additional information and a great deal of encouragement. We also want to thank Dr. C. J. Alley, Professor Emeritus of Enology, University of California, Davis, for his helpful comments on chapter 5; Mr. George Miller of Miller Nurseries for filling the gaps in our knowledge of grape varieties; Dr. Ralph Garren, Extension Specialist at Oregon State University in Corvallis, Oregon, for all of his information on berries and for showing us the experimental brambles; Mr. Ron Reimer for useful pointers about apricots; and Mr. Brian Campbell for his helpful review of chapter 9. David Bilderback, Diane's husband, also played an invaluable role in reviewing the entire book, for which he has earned our undying gratitude. And finally, we want to thank our editor, Suzanne Nelson, for all the hard work she put into this project. Without her comments, questions and coordination efforts, this book would never have come about.

INTRODUCTION

As we set out to write this book our goal was to present information that will help you, our readers, understand how your plants grow. We wanted to make this information on plant growth useful, so that you can see how to work with your plants as you care for them. Once you learn to give them what they need, they will repay you with an abundant, consistent harvest. We want you to be able to understand the "whys" behind the "how tos" of growing fruit, so you can make intelligent decisions about what plants to buy, where to put them in your yard, and how to care for them over the years. Other than strawberries, fruit trees and berry bushes usually represent a substantial investment in both time and money. Nothing is more disappointing than to nurture a tree for years, dreaming of the luscious fruit it will produce, only to have it die during a harsh winter; or to see fruit start to form but never be able to harvest it because it needs more time to ripen and cold weather and frosts are closing in.

We hope that after reading our book you'll be far less likely to make costly mistakes, such as choosing the wrong variety, selecting an inappropriate planting site, or pruning the wrong limbs at the wrong time. By understanding more about your fruit plants, you will not only be able to care for them better and reap more satisfying harvests, but you'll appreciate them more and get a greater sense of pleasure from growing your own sun-ripened fruit.

In this book we limit ourselves to the most popular homegrown fruits. Some of you may be disappointed to find we don't talk about fruits like currants or figs, but we have two reasons for focusing on the most widely grown plants. For one thing, our space is limited, and we want to concentrate on information that will be useful to the largest number of readers. Secondly, the kind of information we provide—about the way plants grow and behave and about the best ways to maximize yield—is available only for crops grown commercially as well as in home gardens. Research money naturally tends to gravitate towards commercially important crops, and some crops are more important than others. This was made apparent to us by the abundance of information we found on apples and peaches, and the lack of data on less commercially important fruits such as apricots.

This emphasis on the needs of commercial growers has shortchanged home fruit growers in other ways, as well. For example, whenever possible we discuss the characteristics, both good and bad, of various rootstocks used for fruit trees. When we know which rootstocks are used by major nurseries, we try to list these. In many cases, there is no truly appropriate rootstock available for a particular fruit tree — the rootstocks that are used have some sort of weakness in disease resistance, hardiness, or some other area. This is not usually the fault of the nurseries; it is due to inadequate research funds and to the great amount of time it takes to investigate a new rootstock thoroughly and find out what its characteristics are. New techniques, such as tissue culture, are speeding up this sort of research significantly, so we can only hope that in the future new rootstocks and new varieties will become available to home fruit growers more rapidly than they have in the past.

We want to point out a very important piece of information we came across in our work on this book: Organic gardeners need to be aware that it is illegal to grow some crops in certain states unless they are chemically sprayed to deter pests. In Michigan, Montana and Washington, for example, cherries must be sprayed to keep fruit flies under control. This law isn't applied just to commercial growers — it applies to *anyone* who is growing a cherry tree. The letter of the law says that a grower who doesn't spray is guilty of "abandoning an orchard." If you want to grow a fruit which is cultivated commercially in your state, you should check with your local extension agent to find out whether there are any laws about spraying.

In order to make the information in this book as useful as possible, we've provided two background chapters. Chapter 1 discusses fruiting plants in general and how they grow and explains what you should be doing to help them flourish and produce a crop. Chapter 6 deals with fruit trees in particular and gives you a broad understanding of their behavior and what sort of cultural attention they require. The rest of the chapters are devoted to individual fruit crops, and this is where you'll find the specific information you need to understand their growth habits and to take care of each crop.

If you're an experienced fruit grower, you may want to skip the general chapters, although they probably contain some tidbits of information that can help you, no matter how many years you've been gardening. Those of you who are just starting out will find that the general chapters give you a good grasp of what fruit growing is all about. At the end of the book we've collected some of the most commonly used terms that pertain to fruit growing and have provided straightforward definitions. If you should come across a term in the book that you don't understand, be sure to check the glossary.

HOW WE HELP YOU CHOOSE VARIETIES

In each individual fruit chapter, we provide you with a guide to varieties that gives information on ripening season, noteworthy characteristics of the fruit, how it can be used, plant growth habit, and zones in which each variety should thrive and produce a crop. It is impossible to list every variety on the market, especially for fruits like apples, so we have tried to pick and choose among what is available—old standbys, heirloom varieties, new improved varieties, and ones that may be grown at the climatic extremes for that type of fruit. These guides are only as good as the information that's available, much of which comes from combing through the catalogs of many nurseries, both for commercial growers and home gardeners. The same variety may be listed as "late" by one catalog and "early" by another; we have gone by "the majority rules" principle when such contradictions occur. When we indicate that a variety is suitable for a certain purpose, such as canning or freezing, we are usually relying on information from catalogs, which is not always accurate. Some books we consulted, especially *Western Fruit, Berries and Nuts,* by Robert L. Stebbins and Lance Wilheim (H. P. Books, 1981) provided less self-serving information, and occasionally we were able to find testing information from state agricultural research projects. You should be aware, too, that some of the comments represent pure, subjective opinion—a grape that one person finds makes a great wine may be considered unfit for that purpose by someone else. In any case, it is wise to use this information only as a general guideline.

In the variety guides, you will see that some varieties have another name given in parentheses; this is because a few varieties go by two aliases—for example, the Italian prune is sometimes known as Fellenberg.

When we talk about disease, we indicate whether a particular variety is susceptible, resistant or tolerant. There's an important distinction between those last two terms. A plant that is bred to be resistant will not be attacked by a particular disease. One that's tolerant will be attacked—but it will be able to withstand the worst effects of the disease and will continue to grow and produce a crop.

When it comes to matching a particular fruit plant to your climate, the United States Department of Agriculture (USDA) has made things a little easier by dividing the United States and Canada into a series of ten hardiness zones. These zones are based upon minimum winter temperatures. In the Vital Statistics box at the beginning of each fruit chapter, we give you the broad range of zones in which each crop can be grown. In the guide to varieties, we give the specific zones for each variety. For some varieties we list an extra marginal zone in parentheses. When you see that, it means that it might be worth a try to grow that particular plant in the milder parts of the marginal zone.

While the zone designations are a handy device, keep in mind that you should use the zone map only as a rough guideline as to which varieties and crops can be grown in your garden. Microclimates can make a very big difference in whether fruiting plants survive the winter or not. If you live at a significantly higher altitude than most of the other areas in your zone, your lowest temperatures will be colder than the "official" minimum ones. The precise location of a plant in your yard can also influence the temperatures to which it is exposed. You may be able to grow a variety successfully which wouldn't otherwise survive in your area if you plant it near the house and give it extra protection in the wintertime. And finally, you must take the pattern of winter temperatures into account. Some fruits such as apricots are very sensitive to fluctuating temperatures and will survive better in an area with

Plant Hardiness Zone Map

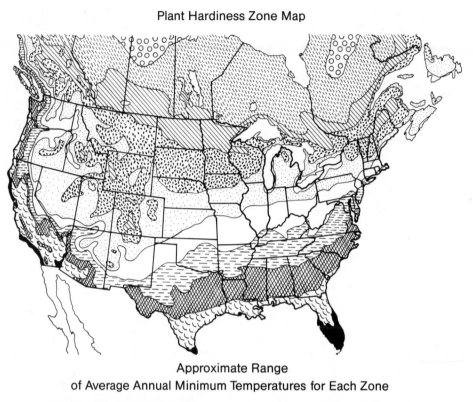

Approximate Range
of Average Annual Minimum Temperatures for Each Zone

Zone 1	Below -50°F		Zone 6	-10° to 0°F
Zone 2	-50° to -40°F		Zone 7	0° to 10°F
Zone 3	-40° to -30°F		Zone 8	10° to 20°F
Zone 4	-30° to -20°F		Zone 9	20° to 30°F
Zone 5	-20° to -10°F	USDA	Zone 10	30° to 40°F

consistently cold temperatures than in one where the mercury bounces up and down like a ping-pong ball.

If you live in the South, of course, you won't be as concerned with winter hardiness. But you should still use the zone designations as a guide, since you don't want to get stuck with a high-chill northern variety.

In addition to cold hardiness, you need to take the length of your growing season into account. Some plants, such as apples and grapes, may survive and bear fruit perfectly well in your area, but the fruit may not reach maturity before cold weather sets in. The information on ripening season we give in the variety guides will let you know whether a particular variety is an early-, mid- or late-season ripener.

Ultimately, the best way to choose varieties for your own garden is to find out what other people in your area are growing successfully and to contact your local extension agent. He or she will be able to provide information about varieties, diseases, soil conditions and just about anything else you need to know in order to make backyard fruit growing a resounding success in your area. Checking with local growers and the extension agent is especially important if you live in an area with a rather quirky climate. If your home is in the Arizona desert or in the Pacific fog belt, you'd better investigate thoroughly before choosing which fruit trees and berry bushes to grow.

Before we started researching and writing this book, we had quite a few seasons' worth of fruit growing experience in Montana under our belts. Even so, we were still baffled by some of the things that had happened in our backyard orchards—why few apricot trees made it through the Missoula winter even though they were supposedly hardy, why Dorothy's Roxbury Russet apple tree never seemed to be able to mature a crop before the weather turned nippy, and why some of our fruit trees gave us a bumper crop one year but nothing the next.

As a result of working on this book, we've solved these mysteries! We now know that we have to look for apricots with high chill requirements as well as apple varieties that will ripen in a short season, and we have learned how to control alternate year bearing with good pruning and thinning practices. These are just a few examples of the enlightening bits of information we came across in the course of our work. We uncovered lots of other helpful tips that we wish we had known when we were first starting out. We offer them to you in the hope that you will learn as much from reading this book as we did from writing it.

CHAPTER 1

FRUITS FOR ALL GARDENS—AND HOW THEY GROW

Are you one of those people who flip through nursery catalogs or gardening books and stop when you come to the pictures of stately apple trees laden with bright red fruit—but then turn the page because you think you just don't have enough room for one in your yard? Or perhaps the picture of a grapevine positively dripping with rich, purple-black clusters sets your mouth watering—but you dismiss the idea of ever growing your own because you're impatient. You don't want to have to spend time tending plants that are going to keep you waiting for years before they give you something to harvest. Maybe you've got the space and the patience, but you think all the pruning and other care the plants require is just too confusing and too hard for you to do.

Well, we've got good news for those of you who have limited growing space, a limited amount of patience, or who may be intimidated by the idea of backyard fruit growing. It's a common misconception that you need a large yard to grow fruit. With the high-yield berries and small fruit trees available today, even a modest garden can be home to a row of strawberries, a decorative and productive fruit tree or a hedge of blueberries, blackberries or raspberries. Now nearly anyone who wants to grow fruit can do so—regardless of the space available. With a large assortment of dwarf varieties currently on the market, you can even treat fruit trees as container plants and grow them in tubs on the patio. That makes it possible for you to take your orchard with you when you move! If you're a lucky landowner with a large expanse of ground, these small-size fruit trees can offer you a benefit you may not have thought of—when you plant semidwarf or smaller trees instead

1

Landscape with a Bonus: Don't overlook the landscaping potential of fruiting trees, bushes and vines. They take no more effort to tend than strictly ornamental plantings, and they give you the bonus of a delicious harvest. Trellised grapes and raspberries make attractive living fences and blueberry bushes make a stately hedge.

of standard trees you'll have a much greater variety of fruit to enjoy since you can grow more of these smaller trees in the same space.

Are you the impatient type? You'll be glad to know that strawberry plants can produce well the same year you put them into the soil; blackberry, raspberry and blueberry bushes as well as some grapevines will start giving you berries the second year after planting; and small-size fruit trees will often bear a harvest two or more years before you'd pick anything from a standard-size tree.

If you have the idea that fruit trees and berry bushes are temperamental and that their upkeep is beyond your gardening skills, take heart. Not so long ago, strawberries couldn't be grown in the South and peaches were too much of a challenge for growers in the Far North. But modern breeding techniques have come to the rescue and have expanded the geographical area in which many of our favorite fruits can be grown. If you've always assumed that you live in an area that's unsuitable for fruit growing, think again. With a little investigation, you might just find that there are varieties able to stand up to whatever your region may have to offer in the way of climatic extremes.

As far as the actual task of growing fruit is concerned, don't be intimidated. Your key to success is to first understand *how* fruiting plants grow, what their special needs are, and then what you can do to help them grow. In this chapter, we're going to talk about things that are common to all fruiting plants, no matter whether it's an apricot tree, a grapevine or a raspberry bush. We'll tell you how they grow and form fruit, how they deal with cold weather, and what sort of general cultural techniques they all require. We'll also demystify the processes of pollination and photosynthesis. This will set the stage for the chapters that follow, which deal with individual fruits. (For general information that pertains specifically to fruit trees, see chapter 6.)

A ROSE IS A STRAWBERRY IS A PEACH

Amazingly enough, almost all the fruits grown in home gardens, from strawberries to apricots, are members of the same plant family, Rosaceae, along with such decorative favorites as roses, mountain ash, flowering quince and potentilla. An outstanding characteristic of the rose family is an abundance of showy flowers. This trait makes most fruiting plants not only productive but also attractive additions to the home landscape. Of the commonly grown fruits we will discuss, only grapes and blueberries are not part of the rose family.

Worldwide, there are about 3,400 members of this very ancient plant group, which exhibit primitive characteristics. If you compare an apple blossom with a daisy (which isn't a family member), you'll see how relatively unsophisticated Rosaceae flowers are. The daisy, while it looks like one flower, is actually made up of many very small flowers clustered together in the yellow center of the blossom. All rose family plants, in contrast, have simple flowers—each blossom is truly a single flower. The flowers usually have five petals, but plant breeders have produced double petals in roses, ornamental plums and some other species. Because they are insect pollinated, the flowers on rose family plants often produce delightful scents, abundant pollen and in some cases, nectar, which attract bees.

The close relationship that so many fruits share has both advantages and disadvantages. On the positive side, it allows nurseries to graft the top of one kind of tree onto the roots of a different, but related species, giving us trees in a variety of sizes and with enhanced disease resistance. The familial bond also makes crossbreeding possible, allowing breeders to create such interesting fruits as plumcots and purple raspberries.

But relatedness also means similar disease susceptibilities, and members of the rose family are all prone to verticillium wilt, a potentially devastating disease which also plagues plants of the solanaceous family such as tomatoes and peppers. (For more on how to protect your plants from this disease, see Problems with Fruiting Plants later in the chapter.) If you're growing any of the ornamental members of the family, disease may spread from them to your fruiting plants. To give you one example, fireblight in your mountain ash could spell doom for a susceptible pear variety in your yard. The same insect pests may bother these related fruits, too; the codling moth and the plum curculio, for example, both munch enthusiastically on a wide variety of popular fruit trees.

ANATOMY OF A PLANT

When you look at a tree or bush, do you ever wonder what's going on underground? Are you curious about how a 10-foot-tall tree is able to move the water it pulls in through its roots all the way to the uppermost leaves? How do fruit trees and grape-vines defy gravity and manage to keep their branches aloft? In this section we'll answer those questions and show you some important things about how your plants function so you'll know what kinds of growing conditions they need to flourish and produce a bumper crop.

ROOTS

The health of any plant, including those that bear fruit, begins underground with vigorous roots. Because they are buried in the soil, we tend to forget about them. But the water and minerals which will eventually become integral parts of the fruit we eat are absorbed from the soil by those all-important roots. Actually, it's the fine network of root hairs that does most of the absorbing. Potassium is the only nutrient that is absorbed through parts of the root other than the root hairs. Blue-berries are the only fruiting crop we discuss that lacks root hairs, and this has important consequences for how you take care of them (as described in chapter 4).

It's interesting to note that the roots of most plants get some help in their nutrient gathering chores from fungi in the soil. The roots interact with certain fungi to produce a mutually beneficial structure called a mycorrhiza. The fungi grow inside or between cells in the plant's roots and extend outside. They help the plant absorb nutrients, especially phosphorus. In soils that have been fumigated to kill nematodes or disease-producing fungi, mycorrhizal fungi are destroyed too, which means that root systems in that soil will be less efficient. The best way to ensure that there are plenty of helpful fungi around is to fortify the planting hole with lots of organic matter. Once the plant is in the ground, keep it surrounded by a thick layer of organic mulch.

In order to do their job effectively, roots need oxygen, which they extract from the air spaces within the soil. If you try to grow a tree in hard-packed or water-logged soil, you won't have much luck since those soils are poor in oxygen and your plant's roots will suffocate! Although some fruit plants will tolerate wet soil better than others, *all* roots need oxygen, so don't tempt fate—be sure to provide good drainage for all fruits you grow.

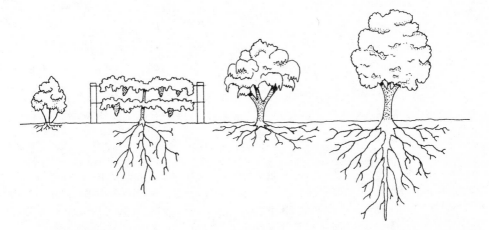

Underground Growth: Some fruiting plants have roots that reach far down into the soil while others stay closer to the surface. From left to right, blueberries stay within the top foot or so of soil, grapes can probe as deep as 12 feet, over 90 percent of a peach tree's roots lie within the top 18 inches and apples send their roots down 10 feet or more.

While the roots of most berry plants are rather fibrous and shallow (within the top 12 to 18 inches of the soil), the roots of grapevines and apple trees can go as far down as 10 to 12 feet if they're in good deep topsoil. For this reason, grapes and apples don't need as much regular watering as most other fruits. Even in a dry area like western Montana where we live, old apple trees survive around abandoned homesteads without any supplementary watering at all. Because of the location of their roots, berry plants and shallow-rooted trees such as peaches need more nutrients in the upper reaches of the soil than do apple trees and grapevines.

In addition to taking up water and nutrients, roots also produce hormones that affect such important processes as shoot growth and cell division within the fruit buds. If the roots of your plants aren't healthy, you can't expect the parts that are above ground to be in very good shape, either.

Like the rest of the plant, roots have growing and resting cycles. In the early spring, a little before the upper portion of the plant shows signs of life, the roots begin to grow. They do most of their growing at this time, then slow down as the shoots above ground develop rapidly during late spring. In the fall, as the top of the plant slows down, the roots begin to grow again. The early start that roots get on their growth is an important reason you should fertilize early in the springtime, so the nutrients will be there when the roots need them.

It's natural to assume that the only place roots can appear is on the underground, anchoring portion of the plant. But the aboveground parts have the potential to form roots, too. If you take a cutting from a grapevine or young fruit tree, you can get it to produce roots under the proper conditions. Roots also have the ability

Mix and Match Plants

Most fruit trees sold today aren't growing on the same roots they started out with. Instead, they're usually the product of at least two different plants—the roots of one and the topgrowth of another—combined by the process of grafting. Without grafting, the wine grape industry in much of the United States would be virtually nonexistent, for the best wine grape plants are very susceptible to a root aphid which kills the vines. To overcome this problem, wine grape shoots are grafted onto resistant roots from native American grapes.

Plants are unlike animals in that they don't have a very sophisticated immune system. When an organ such as a kidney is transplanted from one person into another, the greatest problem doctors face is keeping the body of the recipient from attacking the new kidney as a foreign invader. Plants, however, readily accept tissues from others of their own species, and sometimes even from other species. For more details on how grafting works, see Man-Made Trees in chapter 6.

to produce vegetative growth from dormant buds. In some fruits, these buds frequently sprout, producing shoots called root suckers. Some rootstocks are much more likely to produce suckers than others. While the reasons for these differences aren't fully understood, they probably have something to do with the hormone-producing properties of the different rootstocks. Root suckers are bad news for the plant—they draw energy away from the roots for their own growth, depriving the main part of its fair share. That is why, when we tell you how to prune fruit trees, for example, we always encourage you to prune away suckers as soon as they appear. On other plants, such as brambles, you can use the suckers to start new plants.

STEMS

The stems of fruits come in all sorts of shapes and sizes—from the very short, compact one on a strawberry plant to the tall trunk on a mature cherry tree. No matter what the size or shape, the function is the same. The stem supports the leaf-bearing branches and also transports water and minerals from the roots to the rest of the plant by way of specialized hollow cells which make up the xylem tissue. Sugar manufactured in the leaves during photosynthesis is carried throughout the rest of the plant by way of another type of conducting tissue called phloem.

Both xylem and phloem run through all the roots, stems, branches and leaves of a plant, much like our blood vessels. If a stem or branch is bent and crimped, the conducting tissues may be damaged and not able to carry out their functions efficiently. This is all too easy to do with grapes and with some brambles, as Dorothy has learned with her thornless boysenberries. The canes of these plants are long and thin, and in the process of covering them with leaves in the fall and uncovering them in the spring, Dorothy always ends up irretrievably damaging some. When the canes bend, the xylem and phloem are pinched, and the area beyond the bend dies because no water or nutrients are able to get through.

As we shall see later, under Plants and Cold Weather, stems also serve as storage areas for nutrients produced by the leaves during the growing season. When the leaves turn yellow or red in the fall, that's the sign that the nutrients are being removed to the stems and branches where they will remain until the plant leafs out again in the spring.

GROWTH AREAS

The stem (branches are considered to be stems, too) also contains the growth areas of the plant. The places where leaves are attached are called nodes, while the bare stretches of stem between nodes are called internodes. Plant hormones regulate how long these internodes grow. Sunlight can influence the hormones, so a plant in full sun will have shorter internodes than one that's growing in the shade. If your plant's internodes are excessively long, that might be an indication that it's

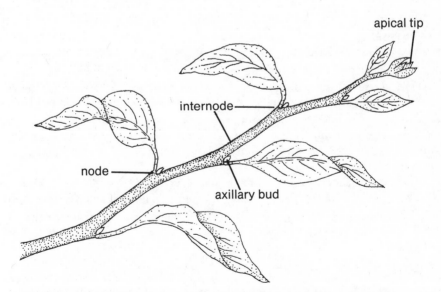

Stem Detail: If you look closely at a growing stem, you'll notice an axillary bud nestled at the base of each leaf. The place where a leaf emerges from the stem is called a node, and the area between two nodes is called an internode. The apical tip at the very end has the potential to form leaves, branches and other plant parts.

not getting enough sun. Lanky internodes are also a sign of too much nitrogen, which would have caused the stems to grow too fast. On a plant that's growing too slowly, the internodes will be excessively short. By noting the length of the internodes, you can get an idea of how healthy your fruit plants are.

The growing end of the stem, called the apical tip, consists of rapidly dividing cells that haven't assumed any particular function yet. As the plant develops, these cells will go on to form leaves, branches and other parts of the plant. The section of shoot with undifferentiated cells is called meristem tissue; this portion of the plant is used for tissue culture. (See New Plants from Old, later in the chapter, for a description of tissue culture.)

At the base of a leaf, right where it joins the stem, is an area of reserve growth tissue called an axillary bud. Hormones produced by the apical tip keep this bud on hold. When the growth of the apical tip slows, the bud is liberated and spends all fall and winter getting ready to leaf out the following spring. Some of these axillary buds will produce leaves each successive season. Sometimes, they can even form long branches, depending on the kind of plant, its state of health, and how much it has been pruned.

The only place where new shoots or leaves can form, other than the tip of the stem, is at the nodes; you'll never see any new growth along the internodes. When you prune your berry bushes, trees or grapevines, always cut the branches off just above a node area. If you leave a long stub, it will die and provide a large area of dead tissue through which diseases and pests can attack the plant.

BRANCHES

The branches of trees and the cordons (branches retained for fruiting) of grapevines harden during the first two or three years of growth then stay rigidly in place. To understand why this hardening occurs, we need to look at cells within the branches. The cells on the top of the branches divide more rapidly and produce more wood than those on the bottom. This extra wood along the top of the branch has more specialized strengthening cells than the underside of the branch does. This reaction wood helps the limbs maintain their position despite gravity and the weight of leaves and fruit.

Once the reaction wood has formed, you can't change the angles of the branches or unbend the cordons without damaging them. It's important for you to spread tree branches and train grapevines while the plants are still young, before the reaction wood has firmly set and dictated their form.

LEAVES

The leaves are the primary energy gatherers of plants. Even though the roots are pulling in water and nutrients, these could not be put to use without the energy provided by the leaves through photosynthesis. (For a closer look at how photosynthesis works, see the following section.) The health and efficiency of the leaves will directly affect the quality and quantity of the fruit, since the energy for their growth and the sugar for their sweetness all come from the efforts of the leaves. You can think of leaves as solar collectors—they must have full access to the sun to be able to collect light energy effectively. Leaves growing in the shade can't even come close to harnessing as much energy in the form of carbohydrates as those growing in the sun can. Much of the thinning, pruning and training that's performed on fruiting plants is done to expose as many of the leaves as possible to the maximum sunlight.

Not only will the health of the leaves affect fruit quality, but it can also influence the survival of the rest of the plant. If a pest or disease causes the leaves to drop prematurely, they won't have had a chance to transport all their carbohydrates to the rest of the plant. As we will see later under Plants and Cold Weather, that stored carbohydrate lowers the freezing temperature of the stem tissues and contributes to the tree's hardiness. Even if a plant that lost its leaves early should make it through the winter, it may still die in the spring after it begins to leaf out. The cause of death would be exhaustion and malnutrition—there simply wouldn't be enough stored food available to make new leaves and continue growth. Diane is concerned because the deer discovered how delicious the leaves of her strawberry plants were late this summer. She's not sure whether the plants were able to store enough food reserves before the leaves were nipped off. She'll just have to wait and see if her plants form new leaves in the spring.

The leaves of a plant can tell you a great deal about its state of health if you know what to look for. Both nutritional deficiencies and diseases can be reflected in the characteristics of the leaves. Small, yellowish leaves, for example, almost always indicate that a plant isn't getting enough nitrogen. We give you more symptoms to be on the lookout for later in the chapter under Nutrients Fruiting Plants Need, as well as in the individual fruit chapters.

We can't stress enough how important it is for you to keep checking the status of your plants throughout the growing season. Far too many gardeners fail to take the time to monitor their plants closely and miss important clues to their state of health. This year, for example, there was a major invasion of the apple leaf roller in our particular corner of Montana. This insect can defoliate a tree by mid-August. While many people were very upset that their apple trees appeared to be dying, a surprising number of them never even bothered to examine the leaves closely enough to see what the problem was. Instead, they phoned the local extension office in a panic, wanting to know what could be wrong with their trees. If they had examined them, they would have seen that the leaves were curled around caterpillars. Don't let a problem sneak up on your plants like these gardeners did. Make it a habit to check the leaves of your plants as you stroll through your yard—they can be an early warning system for serious trouble. If our local gardeners had been keeping an eye on their trees, they could have easily interrupted the life cycle of the leaf roller before it destroyed every single leaf.

MINIATURE FOOD FACTORIES

To understand how the leaves go about producing food for the plant, we need to take a look deep within the structure of a leaf. The vital green color we associate with plants is caused by an energy-trapping chemical called chlorophyll. The chlorophyll is contained within microscopic structures inside the cells called chloro-plasts. Within the chloroplasts, the chlorophyll molecules harness the energy of the sunlight, which the plant uses to make sugar (a form of carbohydrate) from carbon dioxide gas and water. A very important by-product of this carbohydrate production is oxygen. The process of photosynthesis is the fundamental driving force for life on earth. Without the trapping of the sun's energy by plants, other living things would have no source of energy to carry out their activities, and there would be no oxygen for them to breathe (nearly all of the oxygen present in the air is a result of the photosynthetic activities of plants).

As we saw in the previous section, plants draw upon stored carbohydrates to form leaves in the spring. But once the leaves are about half developed, they can start to generate their own carbohydrates through photosynthesis. At first, that energy is channeled into completing the formation of the leaves. But once they're fully developed, the leaves begin to export carbohydrates to the rest of the plant. It's only at that point that the rest of the tree can start to grow and develop.

The amount of energy produced through photosynthesis depends both on temperature and sunlight. A plant subjected to cool, cloudy weather won't generate as much energy as a plant in a warm, sunny location. This explains why the time to ripening is such a hard thing to predict with any accuracy. The same variety of fruit will take longer to develop in a cool or cloudy area than it will in a warm, sunny one at the same latitude. Fortunately for northern growers such as us, our longer days make up somewhat for our cooler temperatures.

THE CRITICAL BALANCE

It's important for you to realize that the fruit you're going to harvest is developing at the same time vegetative growth is occurring. These two vital aspects of the plant's life are competing directly with one another for the plant's energy. When fruit development and vegetative growth are in balance, the plant will produce a quality crop and will put on enough leaves to fuel future growth. But if the plant is carrying too heavy a crop, most of its energy will be diverted to fruit development, and vegetative growth will slow down. To make this idea clear, think of the available food manufactured by the leaves as a pie. If the maturing fruit is using up 60 percent of the pie, then there is only 40 percent left for all the other processes that require energy such as leaf and branch growth, nutrient transport, and the development of new fruit buds for the next year.

Fruit would be much easier to grow if the plants were always in balance naturally, but fruit growers must resign themselves to the fact that this is seldom the case. You should always be ready to step in to establish the proper balance between developing fruit and vegetative growth. Much of the pruning and thinning advice we give is really a guide on how to keep your fruiting plants in healthy balance.

We've already explained how allowing too much fruit to form hinders vegetative growth and the development of fruit buds for the following year. In addition, the harvest itself won't be as sweet since the plant can't produce enough sugar to go around. To top it off, you'll have to wait longer for the fruit to ripen. Because there are so many pieces of fruit, it will take the limited supply of carbohydrates a long time to enter the maturing crop. The ripening process depends on how much carbohydrate has accumulated in the fruit—any delay just pushes back your harvest date. An overabundant crop can also jeopardize the plant's ability to withstand winter's cold, since there will be fewer carbohydrates left over for storage in the woody portions of the plant.

The other extreme, too much vegetative growth, has its own set of dire consequences. If your plants are excessively vegetative, they may not harden off early enough in the fall, which leaves them vulnerable to severe damage if winter sets in prematurely. A plant that's concentrating on putting out an excess amount of vegetative growth will form few fruit buds for the following year. Things like pruning overzealously or adding too much nitrogen can encourage the sort of rampant vegetative growth we're talking about here.

Now that you've seen what happens when you tip the balance to one side or the other, you can understand why we emphasize the importance of keeping growth and fruiting in equilibrium. In the chapters on apples and peaches we're able to give you rather exact guidelines as to how much fruit to allow the tree to bear, depending on the number of leaves the trees has. We can also provide good guidelines for thinning grapes. Unfortunately, we can't be as helpful with the other fruit crops we cover in this book, because horticulturists haven't determined what the optimum fruit-to-leaf-ratios are.

THE TROUBLE WITH THINNING DISTANCES

Many fruit growing guides give what seem to be very precise rules on how much fruit to leave on a tree by telling you how many inches to allow between the pieces of fruit on a branch. Unfortunately, these commonly quoted thinning distances aren't as helpful as you might assume they are. Thinning to distance can still leave a tree with too much fruit if the tree has a large number of fruit-laden branches. And depending on how the fruit is distributed over your tree, thinning to distance can leave a tree with less fruit than it could actually support. Say you have a young peach tree that has a total of 12 peaches, 6 of which happen to fall on the same branch. If you thin by distance, you'll end up leaving only one or two of those on that branch, drastically and unnecessarily reducing your potential crop. If you had sized up the fruit-to-leaf ratio, you might have seen that the tree was well equipped to bear all 12 of those peaches, regardless of the fact that 6 of them appeared on the same branch.

Although they're not perfect, the thinning-to-distance guidelines are all we've got for a lot of crops. Until researchers can give fruit growers more of the helpful fruit-to-leaf ratios for fruiting plants other than peaches and apples, we'll have to use the thinning distance tempered with our own understanding of how plants grow.

If you spend enough time observing your plants, you'll develop a feel for how much to thin in order to keep the fruit and vegetative growth in balance. One year you might count the leaves and fruit on two typical branches just to see what sort of ratio exists. Then, if the tree shows signs that it was bearing too much fruit, you can go back and thin the next year, following a lower fruit-to-leaf ratio.

It's a good idea to look upon each tree, especially when you're growing different varieties, as an individual. Each garden is unique, with its own soil conditions and microclimates. After a few years of experimenting with your fruits, you should become quite adept at keeping your plants in balance.

Diane learned through trial and error that she needs to thin her Italian prune tree more than would be required in a milder climate. Over the years she had come to expect only a few prunes from her tree. But what they lacked in number they made up for in size and taste—they were big and sweet. Last year, her prune crop was unusually abundant, and she didn't thin them enough. That large crop, coming in an especially cool and cloudy year, overtaxed her tree, resulting in a harvest of small, sour prunes that were slow to ripen and didn't store well. Diane knows now that she has to thin the prunes on that particular tree heavily when the crop is large. No doubt you'll have your own tales to tell about lessons you learned the hard way—just make sure you don't forget what you've learned!

FLOWERS AND HOW THEY FORM

Nearly all of the fruit plants we discuss in this book start to form flowers for the following year's crop in the preceding spring or summer. The only exceptions are everbearing strawberries and raspberries, which produce flowers for fall crops in the same year. This is a very important fact to remember. Anything that affects how much food the plant makes, such as diseases, pests, nutrient deficiencies, water stress, severe pruning or too heavy a crop, will generally have a twofold effect; it will influence the number of flower buds created for the following year's crop as well as the quality of the fruit that's forming in the year the problem arises. There are several possible reasons why fewer flower buds appear; one is that the plant is so stressed, it doesn't have the vigor to generate new buds. Or, the vegetative growth may be slowed, resulting in fewer locations for the flowers to form. This latter problem is especially acute in plants such as raspberries and peaches that produce flower buds largely on new wood. When a tree bears too large a crop, it has a built-in means of ensuring the same thing won't happen the next season. Hormones produced by seeds in the developing fruit will inhibit new flower buds from forming. A bumper crop means there'll be a lot more of these hormones around, which explains why fewer flower buds will appear.

Flower buds generally appear at specialized sites that vary depending on the kind of plant. The axil of a leaf—the area where the leaf joins the stem—is the most common location for flower buds. We explain in the appropriate chapters where the fruit buds form on each particular plant. You need to have this information so you can understand how to prune and how to estimate the size of your crop (this can help you thin a tree to bring it out of a biennial bearing cycle).

A QUICK LESSON FLOWER STRUCTURE

Did you ever wonder what went on deep within a flower that transformed it from a pretty blossom into something good to eat? It's a fascinating lesson, and to show you what happens we'll use an apple blossom as an example.

Before the large, open, five-petaled flower unfurls, what you'll see on the tree is a closed bud, surrounded by green structures called sepals. There are generally the same number of sepals as petals. The sepals protect the young flower from frost and insects. Once the apple flower opens up, you can see close to its center a cluster of delicate hairlike anthers, which are laden with yellow pollen. These anthers are the most visible parts of the male portion of the flower, the stamen. In most rose family plants, the flowers produce abundant sticky pollen that adheres to the bodies of bees so it can be carried from flower to flower.

Also in the center of the flower is the pistil, or female part. The tip of the pistil is modified into a sticky surface called the stigma. The area from below the stigma to the top of the ovary is the style. When pollen lands on the stigma, it germinates and

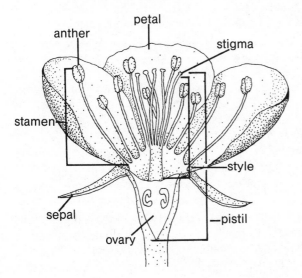

Apple Flower Close Up: If you were to slice a rose family flower like this apple blossom in half, this is what you would see. Each male part, or stamen, is topped with a pollen-laden anther. The female portion, called the pistil, consists of stigmas, a style and an ovary.

develops a long tube that penetrates between the cells of the pistil and grows down to the ovary. Once the tube enters the ovary, a nucleus from the pollen unites with the egg cell nucleus, forming the beginnings of the seed.

This period when the pollen tube is pushing its way down to the ovary is very critical. The temperature must be warm enough, or the pollen tube may grow too slowly. In that case, chances are the egg will be too old to be properly fertilized by the time the pollen reaches it. Or, the point of attachment between the flower and the plant may have already been weakened, so the flower will fall off, even though it was pollinated. When pollination takes place properly, the young seed produces a hormone that sets the developing fruit on the plant so it can't fall off.

While the seed represents a new combination of genetic traits from the two parents, the edible portion of the fruit develops from the tissues of the ovary, which are part of the plant itself. This explains why the fruit you see on one plant is consistently the same—it's all genetically identical. But the seeds within each piece of fruit have the potential to develop into quite different plants, since each has a unique blend of genes.

In order for well-developed fruit to form, a minimum number of seeds must develop. The hormone they produce, called auxin, keeps the fruit on the plant and also promotes the fruit's growth. The more seeds there are within a piece of fruit, the larger it will usually be.

THE MYSTERIES OF POLLINATION EXPLAINED

Have you ever wondered why with some fruits, such as peaches, you can plant just one tree and get a generous crop while with others, such as apples, you need to plant at least two trees of different varieties or you'll get no fruit at all? Trees that can produce a crop without the help of another pollinator variety nearby are said to be self-fruitful. In these varieties, the pollen from the anthers of a flower can successfully pollinate the stigma of the same flower or another flower on the same tree. The pollen from a variety requiring cross-pollination (called a self-unfruitful variety), however, cannot pollinate flowers of the same variety. In nature, cross-pollination is generally the rule, for it encourages genetic variability and prevents inbreeding. Yet in cultivated plants, the need for cross-pollination can be an annoyance. Fruit growers are interested in fruit, not in the new trees that might grow from the seeds. Modern breeders try to develop varieties that will be self-fruitful, but this isn't always possible.

Any one of several problems can make cross-pollination necessary. In some varieties, the pollen won't germinate and grow pollen tubes on stigmas of the same variety. In other cases, the pollen will germinate but the tubes grow so slowly that by the time they reach the ovary the eggs are no longer receptive.

Sometimes, as with Royal Ann and Bing cherries, two different varieties can't successfully pollinate one another because the chemistry just isn't right. Chemicals in the stigma of one variety slow or prevent the growth of pollen from another

Honeybees — Co-workers in the Backyard Orchard

You may wonder how pollen is spread from flower to flower. There's no magic involved — instead, the very mundane forces of gravity, wind and insect activity do the job. Grapevines don't need any insect visitors to carry out pollination, but nearly all rose family flowers plus those on blueberry bushes do need helpful insects around to set a crop. Honeybees are the most common fruit pollinators, but they're rather unreliable workers. They won't forage when temperatures dip below 45°F, and they tend to take time off when the weather is cloudy or rainy. If your insect-pollinated plants have the ill fortune of blooming under these conditions, you should realize that there won't be much pollination going on — so don't expect a record-setting harvest.

Honeybees aren't the most loyal of fruit pollinators, either. If there are any wildflowers blooming at the same time your fruiting plants are in flower, the bees are likely to concentrate on the wildflowers. You shouldn't take this personally — it's just that the nectar of fruit tree flowers isn't as sweet as the nectar of wildflowers. (Even dandelions can be more appealing to bees than apple blossoms!)

If you have only a few flowers blooming on your fruit trees and the bees don't seem to notice them, you can step in. Collect the pollen from one tree with a small paint brush and gently dab it onto the stigmas of the flowers on the other tree. You can also use this method during cold, cloudy weather when the bees tend to stay at home.

The best way to increase the odds that your plants will be pollinated is to increase the number of honeybees in the vicinity of your plants. Many serious home fruit growers raise a hive of their own bees to ensure that they'll have a good crew of pollinators on hand. Then, in addition to harvesting a fine crop of fruit, they also get to harvest jars of golden honey!

You should be aware that insecticide sprays can spell the end for honeybees. If a neighbor happens to spray during blossom time, chances are no flowers will be pollinated since the honeybees will have been killed. No matter whether you have your own hive or are concerned about the welfare of the wild bees, you should try to see that no sprays poisonous to beneficial insects like honeybees are used.

variety that shares certain hereditary traits. If the pollen contains the genes in question, the stigma recognizes this and pollination is thwarted.

In some cases, a variety can't pollinate itself *or* others. This may be because a variety doesn't produce very much pollen, or because the pollen it does produce contains a different number of chromosomes from the variety it is meant to pollinate. For example, Baldwin, Gravenstein, Rhode Island Greening, Stayman and some other apple trees have an extra set of chromosomes. The extra chromosomes lead to faulty, infertile pollen. All these varieties require a neighboring apple tree with normal pollen in order to produce fruit. That second tree, in turn, will require still another variety to pollinate its flowers.

Sometimes, trees that are self-fruitful, such as the sweet cherry variety Stella, will set fruit more abundantly if another variety is planted nearby. The mere presence of more blossoms will attract more bees and increase the odds of better pollination. In certain cases, the pollen from the second variety may germinate better or grow pollen tubes faster than the pollen produced by the first variety. If the weather is cold and pollen tube growth is slowed, that extra edge in terms of speed may result in more flowers being pollinated.

Scientists are a long way from understanding all the reasons for the quirks of fruit flower pollination. Meanwhile, home fruit growers need to follow the advice they are given about planting more than one variety so they can be sure of harvesting reliable crops. When you're choosing varieties for cross-pollination, make sure their blooming periods overlap significantly so they can pollinate one another.

If for some reason you find yourself with only one tree in flower, you can snip a blooming branch from a friend's tree of a compatible variety and hang it in a bucket of water from the branches of your tree. Dorothy did this last year with her Comice pear and was able to harvest a modest crop of pears where otherwise she would have had none.

Should you find that the fruit variety of your dreams needs to be cross-pollinated but your backyard simply won't accommodate more than one tree, consider grafting on a branch of a pollenizer variety, using the T-budding technique we describe in chapter 6 under Doing Your Own Grafting.

WHAT TO DO WITH RIPE FRUIT

All fruits (with the exception of strawberries) have a very ingenious way of pushing themselves into the final stage of ripening. Once they've reached a certain critical size, hormones in the fruit help generate ethylene gas, which in turn triggers the final ripening. Commercial growers often treat fruit with ethylene to promote ripening when the crop has been harvested while still green. Ethylene sets in motion a whole progression of stages in the ripening process including final coloring, flavor development, softening of flesh, and a sharp rise in the respiration rate of the fruit.

Fruit is usually ready to eat just about the time the respiration rate increases. Cold temperatures, such as those in a refrigerator, slow down the respiration rate and help keep fruit from becoming overripe. But you shouldn't look at your regular kitchen refrigerator as the best place to store your harvest. At best, look at it as a short-term storage solution.

The reason most home refrigerators don't make good long-term storage areas is that they're not cold enough! Almost all fruits keep best under the ideal conditions of 30° to 32°F with 90 percent humidity (their high sugar content keeps fruits from freezing within this range). Unfortunately, it's pretty difficult to find a place in a normal home where temperatures remain in this range.

The best solution to the dilemma of long-term fruit storage is to invest in a second-hand refrigerator that you can use just for storing fruit; that way you can turn it down without having to worry about freezing the eggs and the milk. When you stash the fruit in the refrigerator, also set an open pan of water on a shelf to increase the humidity level. Check the thermometer often to make sure the temperature stays where it belongs.

Without these ideal conditions, you really shouldn't count on holding your fresh fruit too long. You're better off eating it right away or freezing, juicing, canning or drying it as soon as possible.

If by chance you're storing apples or pears in a cold cellar or an unheated basement, be careful not to keep other fruit nearby. These two fruits can give off quite a bit of ethylene, and they'll end up hastening the ripening of the other fruit—even to the point where it's overripe! Ethylene can also make potatoes and onions sprout prematurely, so store your vegetables in a different room from your apples and pears.

NEW PLANTS FROM OLD

It doesn't take long to raise a crop of lettuce or beans from seed in your vegetable garden. But if you were to plant a cherry pit or a grape seed, you'd have a *very* long wait before you had anything to bring to the table. Fruit plants can be raised from seed (this is called sexual propagation), but generally only fruit breeders reproduce plants in this way. The other form of propagation, called asexual, is much faster and better suited to home growers since you start with new plants created from stems or roots, and can skip the long, slow process of waking a seed from its dormancy and getting it growing to bearing stage.

Some plants are very eager to reproduce asexually, or vegetatively. Strawberries send out runners, which are actually new plants formed along modified stems of the mother plant. Raspberries renew themselves from buds at the bases of the canes or along the roots. The buds sprout into new canes, which will bear fruit the following year.

When plants are reproduced vegetatively, the new plants are formed from tissues of the old ones. This means the second-generation plants, called clones, are genetically identical to the parent plants. Each variety of fruit, no matter whether it's a strawberry, a blueberry or a peach, consists of genetically identical plants that have been reproduced vegetatively to preserve a distinct set of characteristics.

Plants that don't form clones naturally can still be propagated vegetatively by savvy home fruit growers. Cuttings, divisions, grafting and layering are techniques that can be used on berry bushes, fruit trees or grapevines. In the individual fruit chapters we mention which of these techniques you can use on a particular type of plant. (T-budding, an easy form of grafting, is covered in chapter 6 under Doing Your Own Grafting.)

An exciting, time-saving breakthrough in propagating plants vegetatively is tissue culture. This isn't something you can do at home—the technique is rather sophisticated and calls for sterile laboratory conditions. The way it works is this: A tiny piece of meristem tissue is taken from a plant and placed in a dish of agar; nutrients and hormones are then added to this gelatinous material to spur growth. This tiny bit of tissue starts to grow shoots and roots, and when it's about 1 inch high it is transplanted in a nursery area. The miracle of tissue culture is that an entire new plant can be grown from just a few cells, and in a short span of time. For example, one bud, carved up into 40 to 50 pieces, will provide genetically identical plantlets ready for transplanting in about two to three months. Starting with a ½-inch piece of stem tissue, a nurseryman can produce a 5-foot-tall tree in a year! If he were to use layering or cuttings, it would take him three years to get shoots to a stage where he could transplant them.

PLANTS AND COLD WEATHER

In order to survive the frigid days of winter, plants need to shut down operations until the weather warms up again. Each individual cell in the plant is involved in the process of getting ready for winter. Plant cells are filled with water, and we all know that when water freezes, it expands. If plant cells are frozen before they've had a chance to "winterize," ice crystals will puncture the membranes surrounding the cells, killing them. When plants are resting during the winter, most of the water has been removed from the cells and is stored in the areas between them, so the cells can escape damage.

Carbohydrates, produced by photosynthesis and stored away by the plant, play a very important role in winter hardiness. Some scientists believe that the carbohydrates help pull water from the cells. In addition, they act as a sort of antifreeze— the higher the concentration of carbohydrates in the plant tissues, the lower the freezing point. A plant that's been overpruned or allowed to bear too heavy a crop won't be able to store away many carbohydrates, so it won't be as hardy as one that was properly cared for and has an abundant supply of carbohydrates in its tissues.

HOW PLANTS CAN TELL THE SEASONS

Many processes are involved in a plant's preparations for winter. In a sense, this process actually begins when the plant wakes up and begins to grow in the spring. During the long, warm days of spring and early summer, fruiting plants undergo their most rapid growth. Then they slow down later in the summer. During this lull, most plants are forming flower and leaf buds for the next season.

How does a plant know when to grow fast and when to slow down and begin developing next year's buds? Temperature plays a role for some plants, but that's not the whole answer since the hottest months of the year over North America are July and August, the same months when most plants are already slowing down. A more reliable way of keeping track of the seasons is by measuring the daylength. While July and August are the hot summer months, the longest days of the year are actually in June. June 21, the first day of summer, is the longest day. After that date, the days get progressively shorter until December 21, the shortest day of the year.

Plants have an ingenious way of determining the length of the days. Within the leaves is a pigment called phytochrome. During the day, sunlight changes the phytochrome from one chemical form to another. At night, the process is reversed. The more hours of daylight, the greater the amount of the active form of phyto-chrome present in the leaves. The longer the nights, the greater the proportion of the inactive form. When the ratio of the active to the inactive form of phytochrome indicates that the days are getting shorter, other hormones are produced within the plant which slow the vegetative growth and stimulate the development of buds for the following year.

When frost hits, other changes occur in the plant. Frost triggers the flow of hormones, carbohydrates and other nutrients out of the leaves and into the trunk, stems and/or crown of the plant. The chlorophyll in the leaves breaks down, unmasking the other pigments that were present in the leaves all along—the yellow carotenoids and red anthocyanins. As the green fades away, these other colors come to the fore, giving the leaves their lovely fall appearance. Pear trees and blue-berry bushes are especially eyecatching with their scarlet leaves.

As the hormone auxin flows from the leaf through the petiole which anchors the leaf to the stem, it triggers an important change. Auxin causes a layer of cells to form, called the abscission layer, which will cut the leaf off from the plant. Eventually, the leaf will detach and flutter to the ground. By the time all the leaves have fallen, the plant is dormant and ready to face winter.

SOME POSSIBLE PITFALLS

What happens when something goes wrong with this cycle? A problem that arises all too frequently occurs when diseases, drought, pests or some sort of nutritional deficiency cause a plant to lose most of its leaves prematurely. Then there isn't any way for the plant to sense that the days are getting shorter, for the phytochrome system that counts the hours of daylight is located in the leaves. Warm temperatures or heavy watering can stimulate the plant to put out new growth instead of slowing down the way it should, and a frost could come along and severely damage or even kill this young, tender growth.

As we mentioned earlier, in our city this year there was an epidemic of the apple leaf roller, which feeds on leaves. In addition, the summer was unusually cool and we had a couple of very cold nights in late September. Then, in October, the weather warmed up. This combination of factors totally confused some local apple trees, and they burst into bloom in October! The chances of those trees getting back into synchrony with the climate in time to face winter are very slim. Even if our winter is a mild one, these trees aren't likely to have hardened properly and will probably die.

Another potential disaster for fruiting plants is an early hard frost that hits before the carbohydrates from the leaves have been transferred to the woody parts of the plant. The leaves are killed on the plant and the important food material perishes with them. A plant that's caught like this will have few, if any, carbohydrate reserves, and will be less hardy. Dorothy is concerned because this fall an early hard frost killed the leaves on her grapevines while they were still green. She doesn't know if her plants will survive the winter, or, if they do, whether they will have enough stored carbohydrates to leaf out in the spring. When the weather seems to conspire against you in a situation like this, all you can do is hope for the best.

One peculiar trait of some plants is that, if they are still carrying fruit, an early frost is more likely to affect them adversely than if their crop has already been harvested. This would explain why late varieties of apples and fall-bearing raspberries might be damaged by a frost while earlier bearing varieties that have already been harvested come through unscathed. The reasons for this phenomenon aren't known, but scientists have speculated that once the crop is removed, the hormones needed to promote dormancy become more abundant. This is something to keep in mind if you live in a borderline area for a particular kind of fruit. In such a region, planting a late variety involves the risk that an early frost will hit before you've been able to harvest, and you may lose the plant.

While there's nothing you can do about such assaults of nature as early frosts, there are some things you can do to help your plants get ready for winter. The most important are carefully timing your nitrogen applications (discussed later in this chapter under The Nutrients Fruiting Plants Need) and withholding water in the fall. Also, be sure to control pests and diseases so your plants don't lose their leaves.

DORMANCY AND THE REST PERIOD

Once plants have lost their leaves in the fall, they enter the rest and dormant stages. (Strawberries are the exception—they keep their leaves in winter.) The terms rest period and dormancy are often used interchangeably, but they don't actually mean the same thing. We've tried to make this distinction throughout the book. The dormant phase lasts from the time a plant loses its leaves in the fall until the flower or leaf buds begin to open in the spring. The rest period, on the other hand, is usually shorter. It comprises the time from when the leaves have dropped until the plant has the ability to blossom or unfold its leaves.

At the end of the rest period, a plant will break dormancy if the temperatures are warm enough. If you've ever cut cherry branches in late winter and placed them in water to force them, you've seen how this works. The flowers will open indoors where temperatures are warm, even though the trees outside are still dormant. But if you cut the branches too early, before they've completed their rest period, no amount of warmth will get you a branch full of flowers. The same thing holds true for fruiting plants outdoors—if they haven't finished their rest period, a warm spell won't induce flowering or leafing out. If you plant a fruit variety that requires a long rest period too far south, it won't come to life when it should in the spring, because it won't have undergone enough cold to complete its rest period. It may leaf out weakly later in the season, or it may never leaf out at all.

The length of the rest period varies, both among different fruits and between varieties of the same crop. Those fruits, such as peaches, which originated in milder climes, will generally have a shorter chill requirement than fruits such as apples, which developed in a colder area.

How do plants determine when the rest period is over? Scientists haven't yet completely unravelled the hormonal processes that alert the plant to the passage of hours, but most plants seem to have a way of counting the hours during which the temperature is between 32° and 45°F. When the chill requirement is given for a particular type of fruit it's stated as a number of hours—say 600 hours for one variety of peach and 1,000 hours for a specific kind of apple. Those are hours between 32° and 45°F. When temperatures drop below freezing, the plants don't count them. To complicate matters, certain fruits, such as peaches, may also count hours above 45°F in southern areas.

The whole question of how the chill requirement is met is actually a very complex one. For example, the temperature of the plant itself can be as much as 30°F higher than that of the surrounding air on a sunny day. A few years ago, the temperature in Missoula remained below freezing for four months. There weren't very many cold days before the freeze set in, and spring brought its warm temperatures with a blessed and unusual consistency. During our plants' rest period, there weren't actually very many hours in which the temperature was within the counting range, yet the plants all awakened in the spring in the usual way. What probably happened was that the sunshine during the winter warmed the plant tissues so that they were within the proper range, and hours were counted despite the ambient air temperature.

HARDENING AND DEHARDENING

The whole matter of winter hardiness is a rather complicated one. But if you understand something about what's going on within your fruit plants, you'll at least know what to expect under particular weather conditions.

One complicating factor is that the different parts of the plant may have different degrees of hardiness, and this can vary depending on the time of year. For example, when peach trees first enter dormancy, the buds are hardier than the twigs, so extreme cold is more likely to kill the twigs than the buds. But in late winter, the situation is reversed, and the buds become the most cold sensitive part of the tree.

When the temperature rises above 45°F during the rest period, some fruit plants undergo a dehardening process. This has two repercussions on how the plant comes through the remaining portion of the rest period. When a plant dehardens, it counts backwards, actually subtracting time from the hours it has accumulated to meet the chill requirement. This means that after the temporary warm spell, the plant will need a longer period to meet the chill requirement than it would have if the temperature hadn't risen above 45°F.

The other consequence of dehardening is that the plant becomes, at least temporarily, less hardy. For example, peach buds harden naturally in the fall so that they can survive a temperature of –3°F. If the tree is subjected to winter temperatures of say, 24°F, bud hardiness will increase to –19°F. However, if the temperature rises above 45°F during the rest period, the buds will lose the extra hardiness they gained. They can possibly deharden all the way back to being able to survive temperatures no lower than –3°F, but they won't deharden any more than that unless the rest period is completely over.

Alternate hardening and dehardening can occur throughout the winter in an area with fluctuating temperatures. A tree that was hardy to –30°F in early January might be able to survive temperatures down to only –10°F a month later if there have been a lot of intermittent warm spells. Apricots are especially sensitive to hardening and dehardening. Dorothy learned this when she planted two apricots (Sungold and Moongold) developed by the University of Minnesota that were supposed to withstand temperatures down to –30°F. Both trees died during mild winters in Montana, probably because warm spells alternated with cold ones. The trees dehardened when it was warm and then couldn't take the relatively mild lows, which weren't much below 0°F.

One final irony in the whole hardiness matter is that plants which originated in very cold climates often have a lower chill requirement than those coming from moderate climates! This may seem to be a contradiction, but where winters get very cold and stay that way for a long time, there are few hours between 32° and 45°F for the plants to count. A friend of ours in Montana lost a Carpathian walnut tree, even though it's an extremely hardy tree that can withstand –40°F in the depths of winter. Its weakness is that it has a low chill requirement, so it tends to break dormancy too early in the spring. When it dehardens too soon, it leaves itself susceptible to damage or even death from a late hard freeze.

You can see now why it's important to pick your varieties carefully. If you live in an area where winter temperatures will set the internal clock ticking often in fruit plants, you should be sure to buy varieties with high chill requirements so they won't jump the gun and break dormancy too early.

With all of these complicating factors, it's hard to give clear-cut answers to questions about hardiness. How well your plants can take the cold depends on their health, the weather conditions during a particular winter and even the micro-climate within a specific yard. But it's safe to say that plants that have plenty of stored carbohydrates and are otherwise in good health will survive harsh conditions better than undernourished, sickly ones. Keep in mind that severe cold is not necessarily the worst enemy your plants face—fluctuating temperatures are usually more difficult for them to deal with than consistent cold.

SPRINGTIME DANGERS

The most critical time of year for fruit plants is the end of winter. At this point the chill requirement has been met, so warm weather is all it takes to get the metabolic activities of the plant started again. Once the plant has been reawakened, hardiness is quickly lost—if a late, hard frost should come along, it can spell disaster. This is exactly what happened in our area last spring, when the temperatures dropped in late winter after the plants were reawakening. Many fruit growers lost trees and bushes to that last icy breath of winter. Diane's Canby Thornless raspberries, a hardy Canadian variety, were severely affected by the late cold. Although they didn't die, they were slow to leaf out and some of the canes were blackened and dead on the inside, especially near the tops. With copious quantities of water, Diane was able to help her raspberries compensate for the damage to their xylem and phloem, but her crop wasn't as abundant as it had been in other years.

While scientists don't know exactly what brings about the end of the rest period, they think it has to do with the balance of growth-inhibiting and growth-promoting hormones. One clue is the fact that lots of rain may decrease the chill requirement. Scientists believe that the rain may leach growth-inhibiting hormones out of the plants.

When plants awaken in the spring, they generally grow during the warmer times of day and stop growing when temperatures drop. For this reason, the actual time of blossoming can vary considerably from year to year. If there's a good stretch of warm weather, flowering will occur earlier. Last year, late winter and early spring were so warm that the commercial cherry orchards north of Missoula bloomed two weeks earlier than they ever had before.

CHOOSING A LOCATION FOR YOUR FRUITING PLANTS

You should expect to leave most fruit plants in the same location for years. The only exceptions are strawberry beds, which you should move every few years, and bramble patches, which you can relocate fairly easily. Since you shouldn't disturb all other fruits once you've planted them, it's crucial that you choose their locations with considerable care. When you're sizing up possible sites, keep the future in mind. Don't make the same mistake Diane did when she planted her grapevines. She put them along the south side of the house, which was a good idea, since

What's Wrong with This Location: Can you spot the mistakes the gardener made in locating these fruit trees and blueberry bushes? For starters, they're clustered in a low-lying area that will act as a frost pocket. In addition, their sun exposure will be limited. In the morning they'll be shaded by the evergreens, and in the afternoon they'll be shaded by the house.

grapes need as much of the sun's warmth as they can get. But she also planted a row of aspen trees along the house to the west of the grapes. As the aspens grew, they began shading the grapes. Now, in the summer, the grapes are in the shade from five o'clock in the afternoon until dark, losing valuable hours of additional sunshine. Diane is reminded of her shortsightedness every time she harvests grapes — they aren't as abundant or as sweet as they would be if they were basking in more sun.

Be sure to think ahead before you plant. How tall will other trees in the yard become after ten years? Do you plan to add a garage or other building to your property? Do your neighbors have any trees that could be potential shade throwers? Make sure your plants will receive maximum sunlight, even years down the road.

Where you plant your fruits can mean the difference between success and failure. Even a small yard can have surprising variations in climatic conditions from one part to another. Generally speaking, areas close to the house, especially on the south side, are warmer than other parts of the property, and low-lying areas are colder. Cold air is denser than warm air, so it tends to flow downhill until it's stopped by a building, hedgerow or some other obstruction. This is the reason you should avoid low-lying areas. The best location is partway up a slope, so that the cold air will flow right past the plants before it settles in a frost pocket. One good way to identify where the coldest parts of your yard are is to go outside early on a frosty spring or fall morning and see where the frost has touched and where it hasn't. Those white patches are telltale signs of colder spots that you should avoid.

Not only do temperatures vary within your yard, but they can also differ from one part of town to the next. Diane lives up a narrow, wooded canyon, about 300 feet higher than the airport, where the maximum and minimum temperatures are recorded. By using a maximum-minimum thermometer, she has found that winter minimums are often 5 to 10 degrees colder at her home than at the airport. This means that Diane might fail at growing a fruit variety that thrives only a couple of miles away. This is in fact the case with certain sweet cherries, as Diane has discovered, much to her frustration.

Some of you may live along well-lighted streets. If so, you shouldn't plant your fruit trees or bushes near the street lights. Bright light can fool the plants into behaving as if the days are still long and they won't prepare themselves for winter in time.

If you live in a windy area, you may want to provide some sort of windbreak for your fruit plants. Wind can be quite damaging as it carries small dust particles that will scratch and tear the leaves of your plants, interfering with photosynthesis and decreasing their yield. In one study, it was shown just how important a windbreak can be. Apple trees interplanted with other trees that protected them from the wind were 12 percent larger and gave a 67 percent greater yield than apple trees that were unprotected. If you've got the room, an effective way to shield fruit trees from the wind is to plant a row of fast-growing trees such as poplars on the windward side. Just be sure they're a mature tree's length away so that they don't rob soil nutrients from your fruit trees or shade them when they reach their full height. A hedge or slatted wooden fence also makes a good windbreak for trees and berry bushes.

We have one more word of advice on selecting the best location for your plants. If you live in an area where road salt (sodium chloride) is spread to de-ice roadways and sidewalks in the winter, you should keep your plants well away from these areas, if possible. Salt can harm and even kill fruit trees and berry bushes.

WHAT TO DO WHEN YOUR PLANTS ARRIVE

If you order your fruit plants through the mail from a nursery, you can't be sure they'll arrive when the weather is right for planting or even when you have the time to put them in the ground. Should good fortune smile upon you and enable you to plant them the same day they arrive, unwrap the plants carefully so you don't inadvertently damage the bark or roots. Then, simply soak the roots for a few hours before you put them in the ground, so that they don't suffer water stress.

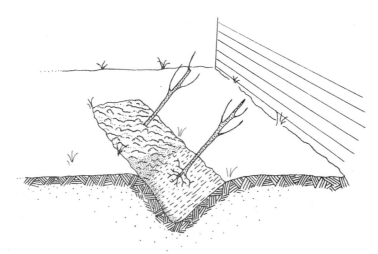

Heeling-In: If you can't plant your fruit trees or berry bushes right away, the best place to hold them over is outdoors in a temporary trench. In a protected area, dig a trench with a sloping side. Lay the plants in the trench, cover their roots with soil and keep them moist.

If you must delay planting, you need to take precautions so the plants don't weaken or even die before you get them into the ground. With strawberry plants, you can put them into the refrigerator in their original shipping wrappers until you plant them. With other fruiting plants, open up the packaging material to check them, even if the wrapping says not to open until you're ready to plant. Plants that have begun to leaf out should be planted as soon as possible. If you just can't get them into their permanent spots right away, you should at least give them temporary lodging in the ground, preferably in a protected location near the house. To heel plants in, as this procedure is called, dig a shallow trench with one sloping side so you can lay the plants down at an angle. Cover the roots completely with soil and water them well. When you're ready to plant, prepare the permanent holes before removing the trees or bushes from the trench.

Plants that are already leafing out can take about one week of being heeled in—if you leave them there any longer you're seriously lessening their chances of surviving the move into their permanent holes. Because these plants have already been "switched" on, they'll start to put out feeder roots into the soil. When you dig the plants from the trench, you'll rip up these tender new roots and the plant will have expended precious energy in vain.

If the plants haven't begun to leaf out, you have two options. You can either heel them in or, if you'll be able to plant them in a couple of days, you can put them in cold storage. Sprinkle water lightly on the packaging material (usually peat moss or wood chips) around the roots and rewrap them. Don't add too much water, or you could encourage damaging fungi to grow. Leave the plants in a cool, protected place such as an unheated garage or basement until a few hours before planting. At that time, unwrap them, trim away any broken roots, set them in a bucket to soak for a few hours, and then plant.

THE NUTRIENTS FRUITING PLANTS NEED

Just like people, plants need good nutrition. A healthy, well-nourished plant will be able to withstand the invasion of disease organisms and pests better than one that's undernourished, and a thriving fruiting plant will produce a far better crop than one that's sickly. A plant growing in humusy soil surrounded by a sea of thick, fluffy organic mulch isn't really going to suffer from many deficiencies. But if one should develop, the trick is to be able to recognize it promptly, so you can treat it without delay. Learning to spot nutritional deficiencies is especially important since all but a nitrogen deficiency can take a long time to overcome. You should be prepared for the fact that a serious deficiency can affect your crop for two years in a row before the condition clears up.

While you're planning your fruit garden, phone the local extension agent to find out if soils in your area tend to be deficient in any particular nutrients. Make every effort to test the soil before planting. If any deficiencies show up, you can include a source of the missing nutrient under the mulch you spread around your new plant and start correcting the problem right away.

One of the nice things about organic fertilizers like compost, manure and rock powders is that they break down slowly and give your plants a steady dose of nutrients over a period of time. It's not the sort of flash-in-the-pan effect you get with rapid-release chemical fertilizers, which all too often wash out of the soil before the plant can use them. There is a lag time before nutrients from organic sources become available to plant roots, and you need to accommodate this in your feeding schedule. If you spot a nutrient deficiency midseason and apply an organic fertilizer, your fruit tree or berry bush still may not make too much headway that season because the nutrients will probably be available for only a short time before the plant shuts down for the winter. Try, whenever possible, to add your fertilizer very early in the season, before the plant breaks dormancy.

When you do add fertilizers, spread them evenly on the soil around the plant, out to the drip-line. In some cases, we're able to give pretty exact guidelines about how much of a certain nutrient to add—but those are really the exceptions. It's tricky to play a numbers game with organic fertilizers and lay down hard-and-fast rules about how much to apply. A material like cow manure can vary in its nutrient content based on the age of the animal, what it was fed and how the manure was handled. Unless we specify otherwise, apply a layer of compost or manure an inch or two deep and a thin dusting of one of the dry powders or meals. It's always safer to start out with a small amount and work your way up in increments, rather than overdo it right from the start, which with certain nutrients can have disastrous consequences.

In this section, we'll explain what the important nutrients are, what they do for plants and what the universal deficiency symptoms look like. (When a plant exhibits a special set of symptoms, we'll mention them in the chapter on that plant.) We can't stress enough how important it is for you to keep checking your plants all season so you can head off any serious nutrient shortages before they drag on for too long. Along with our discussion of each nutrient we'll mention some commonly available organic sources.

NITROGEN

Nitrogen is the number one nutrient for fruiting plants. They need it to form genetic material, plant hormones, proteins and chlorophyll. The most universal symptom of nitrogen deficiency is yellow or yellow-green leaves, which are due to a scarcity of green chlorophyll. The leaves take on the color of the underlying yellow carotinoid pigments as the chlorophyll subsides. Plants starving for nitrogen also grow slowly and have smaller than normal leaves. This makes sense, for without adequate chlorophyll, a plant is weak in energy-capturing power and can't grow very fast.

Nitrogen is the one nutrient that you must give your plants every season after they've been in the ground for a year or two. For some fruits, we're able to give recommendations as to how much nitrogen you should add each year. We can't do this for all of them because the appropriate research hasn't been carried out. These recommendations are made with an average loamy soil in mind. Heavy clay soils or soils rich in organic matter will hold nutrients longer and will need less nitrogen than we recommend, while sandy soils, which act almost like sieves when water and nutrients are applied, will need more.

To figure out how much fertilizer to add to provide the target amount of nitrogen, you need to know the approximate percentage of nitrogen in your source. When you buy fertilizer such as bagged manure or blood meal this information is usually given. For example, steer manure usually has about 0.7 percent nitrogen. In a 100-pound bag, there will be 0.7 pounds of actual nitrogen. Rabbit manure contains 2.4 percent, pig manure 0.5 percent and poultry manure about 1.1 percent. Blood meal has 15.0 percent nitrogen, cottonseed meal 7.0 percent and fish emulsion 5.0 percent. Once you know the percentage of nitrogen in your source and the amount of actual nitrogen needed by your tree or bush, you can calculate how much to use.

You really shouldn't make a habit of giving nitrogen to your plants beyond early spring. Because of the slow-release rate, an organic fertilizer you applied in midsummer could release its nitrogen in fall—a time when any new growth the plant puts out will interfere with its hardiness. However, if a deficiency does show up during the growing season, that doesn't mean you should ignore it. There is a way you can add nitrogen—give it in a quick-acting form such as fish emulsion or manure tea, and don't apply it too late in the season.

With fruit trees, you're not likely to see the effects of an extra dose of nitrogen in the same season you add it. The early growth, flowering and fruit set of fruit trees are dependent on nitrogen stored in the bark the previous year. So, even if you add nitrogen early in the spring, a tree that has been deficient won't respond until the following year. Along the same lines, a tree that had too much nitrogen the previous year will still grow too much in the spring, even if you withhold nitrogen.

PHOSPHORUS

Plants use phosphorus to help make the carbohydrates during photosynthesis, to produce genetic material, and to carry out a host of other metabolic processes.

A plant that's low on phosphorus will signal this by the color of its stems and leaves. The stems may appear purple and the young leaves will be unusually small and darker colored than normal. As they mature, the leaves will become mottled and occasionally bronzed.

All fruiting plants take up phosphorus from the soil with the help of mycorrhizal fungi. The trees or bushes you buy were probably grown in nurseries where the soil was fumigated, which means the mycorrhizal fungi were killed off. Once these plants are in your possession, you can make up for their bad start in life. Soil that's chock full of humus, and a fluffy organic mulch, will provide plenty of fungi to help your plants harvest phosphorus from the soil.

Most fruit plants will need little if any added phosphorus as long as they're mulched and/or fertilized with manure. But if a deficiency should occur, bone meal, phosphate rock and wood ashes are three reliable sources. Be aware, however, that wood ashes used in large quantities can raise the pH of the soil. When you're looking around for bone meal, try to find the steamed kind rather than the ground type. Steamed bone meal is broken down into smaller particles, so it releases its phosphorus more rapidly than ground bone meal.

POTASSIUM

Potassium keeps plant enzymes active and helps the leaves open and close their pores to take in carbon dioxide and to let out moisture. This nutrient also plays a vital role in the transport of important materials in the xylem and phloem. Plants use lots of potassium and may need more than is present in the soil. Even if there's enough in the soil, young trees may not be able to make full use of it, since their root systems aren't very extensive. Remember, potassium is taken up by the roots themselves, not by the root hairs. Since the roots don't have as much total surface area as the network of root hairs does, they really aren't as efficient at gathering nutrients.

Your tree is telling you it's low on potassium if the leaves are bronzed or look scorched along the edges. Sometimes they will also cup or roll. A slowdown in growth is another sign that an infusion of potassium is needed. Manure is a good potassium source, along with granite dust and greensand (mined from undersea deposits). If your soil is too acid as well as lacking in potassium, an application of wood ashes would be a good way to clear up both problems.

MICRONUTRIENTS

Boron's most important job is to help move sugars via the phloem. Plants without enough boron often develop rosettes of leaves. That's because the internodes are so shortened by poor growth that the leaves all appear to come out from the same area of the branch. The leaves themselves may even be deformed. Another sign that more boron is needed is a reluctance to break dormancy. The soil in some regions tends to provide less than adequate amounts of boron. This is the sort of information your local extension agent can give you. Acid soil and abundant rainfall can also lead to a boron deficiency.

Organic matter is the best long-term source of boron. If your soil is low on humus and a soil test reveals a boron deficiency, you can add phosphate rock to overcome the shortage. Another short-term solution would be to add some boron in the form of household borax. Scratch 4 to 5 ounces per 1,000 square feet into the top inch or so of soil. You should only resort to this in the face of a serious deficiency. Unfortunately, it's easy to overdose with boron, and too much can lead to oozing sap or dying leaves. So play it safe and always be conservative in adding boron.

Plants need iron so they can form the proteins they use in photosynthesis and respiration. Iron is also an important link in the chain of chemical events that make nitrogen available to plants. The symptoms for an iron and a nitrogen deficiency are similar, but with careful observation you should be able to tell them apart. The leaves of iron-deficient plants often turn yellow, but retain green veins, whereas nitrogen-deficient leaves are uniformly yellow.

If your plants look like they need iron, you should test the soil pH, for alkaline soil and an iron deficiency often go hand in hand. When the pH is too high, the soil holds iron in an insoluble form, so even though it's there the roots can't use it. Often, correcting the pH will eliminate the iron problem as well. If the pH is normal, consider adding glauconite or greensand, two materials that supply iron.

Magnesium is important in helping plants respire and it's an essential component of chlorophyll molecules. This micronutrient also aids plants in capturing the nitrogen and phosphorus that are in the soil.

The sign that a plant is low on magnesium is lower leaves that turn yellow (except for the veins, which remain green). Leaf edges show the changes first—they turn yellow, then orange and finally brown. Eventually the whole leaf may die. In a serious deficiency, many leaves die and fall prematurely. Well-mulched plants are not likely to show a magnesium deficiency, but if by some chance they do, give them some dolomitic limestone. This is the same material you use to correct an overly acid soil, so keep in mind that too much may start to have an effect on your soil's pH.

Zinc is needed for the formation of the hormone auxin, which is an important growth promoter in all parts of the plant. Zinc-deficient plants will have rosettes of leaves, and the leaves are likely to be abnormally small. They may also be partly or completely yellow.

Out of all the micronutrients, zinc is the one that plants are usually lacking. A soil that's low in organic matter or one that's compacted and has a hardpan layer will very often set the stage for a zinc deficiency. Manure is the best source of zinc, and phosphate rock is a close second. Cornstalks and hickory, peach and poplar leaves all accumulate zinc and are good materials to add to the mulch around a zinc-deficient tree.

POINTERS ON GENERAL CARE

Fruit plants really aren't as finicky as you might think. In this section, we'll run down some general instructions on how to care for them and point out ways to avoid possible problems. In each chapter on an individual fruit plant, when it's appropriate, we'll give you more specific information.

MULCHING

No matter what fruit crop you're growing, it's worth your while to establish a permanent mulch system around the plants. The mulch will conserve moisture, allowing the roots to grow in the richer surface layers of the soil, where nutrients are more abundant. Only in moist soil can roots absorb nutrients. In addition, the mulch itself will add a broad spectrum of nutrients as it breaks down.

Compost and hay make good mulches, but avoid hay that has large amounts of legumes like vetch and alfalfa. Because these materials tend to break down slowly, they release their nitrogen late in the season, right when plant growth should be slowing down, not stimulated.

When you lay down a hay mulch, fortify it with manure. This will ensure that the soil microorganisms that are breaking down the hay won't steal nitrogen from your growing plants to fuel their decomposition activities. The manure will supply an extra dose of nitrogen so that both the plants and microorganisms will have their share.

Make sure your layer of mulch stays 3 to 6 inches deep. You'll probably have to add more material each spring. By keeping this layer nice and thick, you're discouraging weeds and saving yourself some work. Make sure the ring of mulch doesn't touch the base of the plant—leave 2 to 4 inches of breathing room so there's no chance of rot or disease developing.

WATERING

It's difficult to give firm guidelines about watering fruit plants because there are so many variables involved. Various soil types dry out faster than others and certain plants with deeper roots need less water. As a starting point, you should be prepared to give plants a deep, thorough soaking at least once a week. Plants in clay soil will need less watering than those in loamy soil, and plants in loamy soil will need less than those in sandy soil. Weather can also be a factor—during a hot, dry spell you'll have to increase the frequency.

The best thing you can do is to really be in tune with your plants and watch for the signs that they need more water than you're giving them. Berries will often signal this with dull-looking leaves or, if the water stress is bad enough, drooping leaves. But with fruit trees and grapevines, your only clue may be very slow growth. Don't wait until you see wilting leaves—when a tree or vine reaches that point, it's almost dead.

Be sure to give your plants enough water while they're blooming and forming fruit. If you're negligent, they may drop their blossoms, the fruit may not set well, and the size and quality of the fruit that does form will be inferior. Certain plants like grapevines, apple trees and pear trees will form fewer flowers for the following year if they're under water stress during the period when that essential flower bud formation is going on. Plants that are water stressed are also more prone to diseases and pests than ones that are well watered.

Too much water can be as much of a curse as too little. Don't oversaturate the soil or you run the risk of suffocating your plants' roots. Water fills the air spaces in the soil pores, and your plants need both water and air to live. No matter whether you're using an overhead sprinkler, drip irrigation system or hand-held garden hose, if the water puddles on the soil surface, you're giving too much too fast. The key to good watering is a slow, steady soaking.

You should alter your watering routine to allow plants to prepare themselves for winter. For starters, ease up on the frequency of watering as fall sets in. In late fall, give your plants one final, deep watering before the ground freezes, unless it's been raining heavily. Even a dormant plant needs some moisture to survive the winter.

ATTENDING TO SOIL pH

Most fruiting plants do well in a soil that's between 6.5 and 7.0 on the pH scale. (Blueberries are the notable exception—they take much more acid soil as we'll explain in chapter 4.) You should check the pH at least once, preferably before planting, and if your plants keep on growing well, there's no need to recheck it. When a problem develops and you're trying to diagnose it, you'll probably want to test the pH again. Very often when the pH has crept outside the optimal growing range, nutrient deficiencies ensue.

PROBLEMS WITH FRUITING PLANTS

The best way to handle pests and diseases is to use the equivalent of preventive medicine in the backyard orchard. If you know a particular pest or disease is a problem in your area, buy resistant varieties if they're available. Check your plants often for suspicious insects and disease damage during the growing season so you can take care of a problem before it gets serious. Remove all rotting or dropped fruit regularly so you don't give pests or diseases a place to multiply. Good pruning practices will promote good air circulation which will discourage fungus diseases. Healthy plants that receive good light, ample nutrients and plenty of water will be more resistant to pests and diseases than neglected, unhealthy plants. When you plant a fruit tree, berry bush or grapevine, it's like adopting a child—in either case, neglect is abuse.

In the chapters on individual fruiting plants we'll call attention to the various problems that may befall the different types. There is one disease that does strike *all* rose family members though, and that's verticillium wilt. Plants in the solanaceous family such as eggplants, peppers, potatoes and tomatoes are also susceptible to this fungus disease. When you plant any rose family fruits (and this includes trees), avoid sites where other susceptible crops were grown before. Even if they never exhibited any symptoms of the disease, they could still have been carriers and the fungus might be lurking in the soil. One commercial grower who didn't know this invited economic ruin by planting 50 acres of black raspberries where potatoes had grown before. During the first summer the raspberries grew well, but in the fall the entire crop was destroyed by verticillium wilt. The potatoes had been resistant to the disease but had secretly harbored it. When the raspberries were planted, the wilt organisms in the soil took over.

Heavy, poorly drained soil seems to foster this fungus disease, so an important preventive measure is making sure you plant in crumbly, well-drained soil. Also, be sure you start out with plants that are certified to be disease free. In the individual fruit chapters we give the symptoms to watch for. If your plants should happen to come down with verticillium wilt, be sure to remove them immediately, roots and all (this goes for trees, too).

For solving problems with pests and diseases beyond those discussed in this book, *The Encyclopedia of Natural Insect and Disease Control,* edited by Roger B. Yepsen, Jr. (Rodale Press, 1984), is a good reference. You can also call or write to your local extension agent to find out what "enemy" you're facing; just be prepared for the fact that he or she is likely to tell you how to solve the problem with nonorganic means. But at least you'll have identified what you're up against, and you can then turn to an organic guide to see how to deal with the problem safely but effectively. If you have a really difficult problem, writing to the editor or question and answer column of a gardening magazine can sometimes bring the help you need.

Some areas of North America suffer from a problem that's neither a pest nor a disease—salty soil. The soil has accumulated excess salt in regions where rainfall is scarce (so there's little natural leaching of salts from the soil) or where salty irrigation water or chemical fertilizers have been used over the years. Plants don't do very well in salty soil. They develop brown regions at the leaf margins when there's only a moderate amount of salt. If the salt concentration is high, the leaves turn completely brown and fall off prematurely.

Fruit plants vary in their sensitivity to excess salt. Strawberries and raspberries are the most sensitive, followed by blackberries, apricots, peaches, plums, apples and pears. Grapes are the most tolerant of salty soil, but even they are not very accepting of an excess amount of salt.

If you suspect that your soil suffers from this problem, you can have it tested. Call your local extension agent for information about collecting samples and having them tested. If it turns out you do have a problem and your soil is reasonably well-drained, you can sometimes leach the salt out by watering with copious amounts of unsalty water around the roots of your trees. You can plant strawberries and brambles in 6- to 12-inch-high hills of fresh, non-salty topsoil you've imported from elsewhere and keep the soil moist to help minimize the salt content of the soil near the roots.

CHAPTER 2

STRAWBERRIES
SPRING'S FIRST FRUIT

Vital Statistics

FAMILY
Rosaceae

SPECIES
Fragaria × *Ananassa* (common garden
strawberry)
F. vesca (Alpine strawberry)

POLLINATION NEEDS
Self-pollinating

HOW LONG TO BEARING
Everbearers produce towards end of
first summer following planting. Junebearers produce in second summer
following planting. In Deep South, fall-
planted Junebearers produce in early
spring following planting.

CLIMATE RANGE
Throughout all of the United States and
lower Canada

The bright red strawberry is a fruit of many virtues—besides possessing a superb flavor it is inexpensive, prolific, and widely adaptable. Best of all, it produces heavily a year or even less after planting, long before other fruit trees or berry bushes are anywhere near ready for a first harvest. If you're an impatient, hungry gardener, strawberries are the perfect crop for you! And as if all this weren't enough, strawberries can be grown in the smallest of gardens, in patio planters or in hanging baskets.

Despite their many attributes, strawberries will only do well if you know how to choose the right varieties, plant them correctly and take care of them properly. Varieties that grow well in the northern part of the country will fail in the South, while southern strawberries will freeze to death in the chilly northern autumn. Although strawberries are tough plants that can take some abuse, if you plant them improperly they'll die before they even have a chance to grow at all. And if you plant helter-skelter without planning your strawberry bed carefully, the bed may bear bountifully for a couple of years, only to decline the third year to practically nothing. But don't let all this discourage you! Strawberry growing really isn't tricky once you know something about these plants and how they behave.

SOME STRAWBERRY HISTORY

A good place to find important clues to the strawberry's behavior in the garden is to examine its history and biology. Strawberries, it seems, have been around for a long time. Prehistoric people probably dined on the tiny wild strawberries that still grow in the wooded areas of many parts of the world. The Romans were the first people on record to actually cultivate these plants. These early growers did such a good job that the species they grew, *Fragaria vesca,* has remained popular in Europe ever since. Improved varieties that we know as alpine strawberries or *fraises des bois* were developed in France during the nineteenth century. (We'll talk more about these later in the box on Strawberries from Seed.)

Another European strawberry, *Fragaria moschata,* has separate male and female plants. This berry was cultivated in the eighteenth and nineteenth centuries under such intriguing varietal names as Black Strawberry, Apricot Strawberry and Raspberry Strawberry. Unfortunately, these varieties have completely disappeared from the scene and we don't have a hint of what they looked or tasted like.

The garden strawberry of today is actually a hybrid produced from two American species, *Fragaria chiloensis* and *F. virginiana.* The first, *F. chiloensis,* lives along the western coasts of South and North America, ranging as far north as California. In Chile, it not only grows along the frost-free coast, but also up into the Andes Mountains, where temperatures can drop to $-30°F$! The other parent species, *F. virginiana,* ranges over much of the United States, from the bitter cold snowfields of Alaska to the warm and moist deltas of Louisiana.

As you can see, the difference between the native climates of these parent species is quite dramatic, and helps account for the great variability in climatic tolerance among their offspring. As long as the variety is properly chosen, strawberries can be grown anywhere from the tropics to the arctic. There aren't too many other fruits for which we can make that claim!

Ironically, the breeding work between these two diverse American species that led to the modern hybrid was done in Europe. This interesting tale starts off in the eighteenth century with a bit of intrigue. It seems that the Chilean strawberry was brought to Europe in the early 1700s by a French spy. (What the spy was doing with these plants in the first place remains a mystery.) This species has separate male and female plants, but unfortunately, only female ones were brought ·

to the continent. After some unsuccessful attempts to get fruit, people realized what the problem was and began to grow the new strawberry in alternate rows with two other species: the European *Fragaria moschata* and the American *F. virginiana,* which had been brought to Europe at an earlier date. Hybrids developed between the two American species, and luckily they inherited the perfect flowers of the Virginian species. The first hybrids were described as a separate species, *F.* × *Ananassa,* in the mid-eighteenth century. These must have been interesting berries, for the fruit was pink and had a flavor somewhat reminiscent of pineapple.

Over the ensuing years, more varieties were developed, including everbearing types. The tremendous genetic variability of the strawberry was utilized in the breeding activity throughout the nineteenth century to produce fruits with such tantalizing flavors as apple, apricot, cherry, grape, mulberry and raspberry. Unfortunately, we'll never be able to taste them since all these types have been lost over time. Since the original wild species still exist, however, some curious breeder might be able to recreate these fruits for twentieth-century gardeners.

EVERBEARERS AND JUNEBEARERS — WHICH KIND TO GROW?

Strawberries come in two basic types—the Junebearers, which produce one crop in June or July, and the everbearers, which give you two crops, one in June or July and another, smaller one in late summer or early fall.

Faced with this choice, how do you decide which one is the best for you to grow? Actually, it all boils down to two considerations: your plans for the crop and your geographic location. Gardeners who expect to freeze most of their berries or turn them into jam or preserves will want to grow Junebearers for their concentrated crop. Fresh strawberry lovers may prefer the longer harvest period of the everbearers, if these plants will grow and produce well in their area. If you understand the factors that affect strawberry flowering and fruiting, you'll be better able to make the decision about which berries will do best in your garden.

Although strawberries bloom in the spring, the flowers you see actually began developing in the late summer or early fall of the previous year. Cool temperatures accompanied by days of 12 to 14 hours initiate the formation of the embryonic flower buds. Then the plants become dormant for the winter. This chilling period is necessary to stimulate the production of plant hormones that will induce the flowers to develop further. Come spring, dormancy is broken by the warmer temperatures and the flowers complete their development and bloom. By the time the first clusters of flowers have developed into fruit, it is summer, the days are long, and the temperatures are warm. In Junebearing varieties, these factors inhibit the plants from continuing with the process of flower formation until the shorter days and cooler nights of fall come again. This is why Junebearers give most home gardeners only one crop of fruit each year.

Commercial growers in southern California use this on-off cycle of flower formation to their advantage. They buy strawberry plants grown at high elevation or more northern nurseries and plant them in the beginning of November. The plants begin to bear as early as February, and, because the days that follow are short and cool, the plants keep producing abundant flowers and berries. The picking doesn't stop until the end of June when warm temperatures and long days call a halt to flowering.

In general, Junebearers respond to long days by initiating runners. Either during the blooming period or right after, when the days are warm and around 15 hours long, the first runners appear. Those early runners will ultimately produce the most fruit the following year.

Everbearers respond differently to daylength—they keep producing flowers under long day conditions, and in fact, the longer the days, the more flowers they will initiate. For example, horticultural researchers have found that the variety Geneva will form an average of 8.8 flower clusters per crown when days are 18 hours long and only 1.4 during 12-hour days. Because everbearers produce the heaviest second crops during long days, they can be disappointing to gardeners in areas where summer days are shorter than 15 hours. But in more northerly parts of the country with nice long days, these adaptable varieties will produce lots of fruit.

Because they're devoting their energies to berry production, everbearers don't make as many runners as do Junebearers. While enthusiastic Junebearing varieties such as Catskill may produce an average of seven runners per plant in one season, each plant of conservative everbearers like Geneva may make only four runners in a year.

Daylength and temperature can sometimes play tricks on plants, and jar them out of their flowering habit. Some varieties that are everbearers in Oregon and northern California may become Junebearers in other regions. Dorothy bought some plants of the variety Sequoia, which was listed in the catalog as an everbearer. After the first luscious crop was harvested, she waited and waited for the second set of blossoms to come, but they never did. Because of the long, warm, Montana summer days, the plants stopped making flowers and switched to runner production instead. Sequoia apparently responds to daylength and temperature like a Junebearer, although it fruits like an everbearer in the cool temperatures and moderate daylengths of central California. This is true of some other Junebearers—cool temperatures are key to flowering. Still other Junebearers are more dependent on daylength and won't initiate flowers under longer days no matter what the temperature.

STRAWBERRY VARIETIES

To help you make sure that the variety you select is suited to the conditions in your particular region, here are some general words of advice. (For more guidance, be sure

A Bounty of Berries: If you think that all strawberries look alike, look again. Starting at the top left and moving clockwise, you can see that some varieties are cone shaped, some look like compact wedges, others are a cross between a cone and a globe, some have white necks and certain varieties are shaped like small red globes.

to turn to the Guide to Strawberry Varieties at the end of the chapter.) Everbearers thrive only in areas with long summer days; they won't do well too far north or at too high an altitude, since early frosts will interfere with the second crop in these regions. If you garden in the South or in southern California, you can buy plants each year for an extended late winter and spring crop. Just be sure to avoid varieties such as Fresno and Shasta, which require a long chilling period and won't grow well if fall planted. Tioga and Sequoia, however, do well under fall planting conditions in California, while Missionary and Florida 90 are good varieties for the South.

No matter how appealing a variety sounds, if it's a southern variety and you're a northern grower, you'd be better off finding another. Southern varieties continue to grow vegetatively far into the fall and would be killed off in the North by frost. Winter-hardy varieties, on the other hand, slow down their growth in autumn in preparation for winter. These sturdy varieties, such as Catskill, Sparkle and Sunrise, won't produce under the short days of winter even if they're brought into a greenhouse.

As you read catalog descriptions, keep your eyes open for varieties with bred-in disease resistance. There are even some that tolerate high humidity levels. Any of these characteristics will help keep your strawberry patch problem-free.

If you live in an area with a long growing season, you may want to choose an assortment of varieties with different blossoming and bearing times so that your harvest is spread out. One note of caution, though: in areas where late spring frosts are a threat, only strawberries with later flowering times will be reliable.

Once you've narrowed down your choices based on these factors, consider how you're most likely to use the berries. Do you want to stock up the freezer, or are you strictly a fresh berry connoisseur? Some varieties are better for freezing, while others provide better fresh fruit. (The Guide to Strawberry Varieties will give you an idea of which ones these are.) If you're a jam maker, you should know that some berries make better jam than others. For example, a batch made from Tioga strawberries loses its color more quickly than jam made from Hood berries.

THE FOOD VALUE OF STRAWBERRIES

The main nutritional contribution strawberries make to our diet is vitamin C. The C content ranges from 39 to 89 milligrams per hundred grams of fruit, depending on variety and growing conditions. Aiko, Aptos, Catskill, Hecker, Marshall, Redheart and Robinson are high in vitamin C, while Aberdeen and Blakemore are on the low side.

Vitamin C content and sugar accumulation are both favored by long, sunny days and cool nights. If your strawberries should ripen during a period of cloudy weather with warm nights, your crop just won't taste as good as it would if it had ripened when skies were clear. There's bad news on the nutritional front, too: the vitamin C content will be reduced.

Individual berries that ripen in the sun will be much more flavorful and nutritious than those that mature hidden under the leaves. This is unfortunate for those of us who compete with the birds for our strawberries—unless we cover our plants carefully with netting, we can end up with the less satisfying, concealed berries while the birds feast on the sweeter, more obvious fruit.

Those of you who are calorie conscious will be relieved to know that even though they taste sinfully sweet, strawberries are blessedly low in calories. One cup of fresh berries has only 55 calories—a veritable bargain considering how sweet and juicy they are.

HOW STRAWBERRIES GROW

It's important for gardeners to understand how strawberry plants grow, for their unique growth habits require special treatment. While most plants have an elongated stem with leaves growing from nodes spaced along it, the strawberry plant has a very compact stem with the nodes crowded close together. When you look at a strawberry plant, you see the leaves all originating close to the ground, from a thick central area called the crown. This crown is actually the stem of the plant.

Strawberries are also peculiar in the way the roots form—they grow from the bases of the leaves. For each leaf the plant produces, up to six new roots may also develop. If the roots are damaged or dry out, no more roots can form until new leaves are produced. This growth pattern can cause serious problems for the plant. Badly damaged roots mean that it can't get enough nutrients to grow more leaves. But without new leaves, it can't grow any more roots either! This explains why strawberry planting instructions always emphasize placing your plants so none of the roots are exposed above the ground.

It's equally important not to bury the crown. Remember that the crown is actually the entire stem, and that all the leaves emerge from it. If the crown is buried, the leaves cannot properly grow and expand. A buried crown also encourages

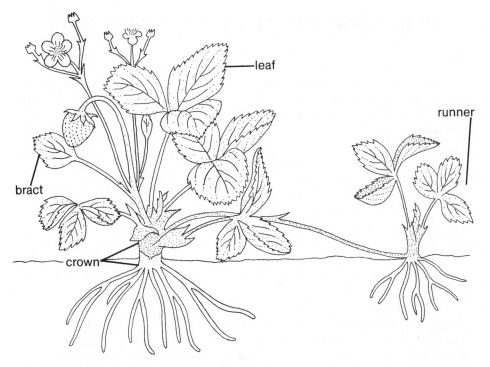

Like Mother, Like Daughter: The strawberry mother plant on the left has sent out a runner. Once this offspring becomes established, it will produce its own bracts, flowers, berries and runners. The thickened area, called the crown, is the plant's stem.

fungal and bacterial activity which can eventually rot and kill the plant. (For more details on finding the correct planting depth, see the section on Planting Your Strawberries later in the chapter.)

LEAVES

Strawberry leaves form in a spiral pattern around the growing point of the crown. If you carefully examine a strawberry plant from above, you'll see its low rosette form. The hybrid nature of the cultivated strawberry is evident in the leaves, which are intermediate in thickness between the thick leaves of *Fragaria chiloensis* and the thinner ones of *F. virginiana*. Some varieties resemble one parent more closely than the other, so the leaf thickness of varieties differs considerably. Chilean strawberry plants stay green and continue to grow through the winter, while Virginian ones die back and take a winter rest. Northern strawberry varieties follow a pattern more like the Virginian species, while varieties which thrive in the South are more like their Chilean parent. The leaves of most garden strawberries have three leaflets like the Virginian plants, but a few which are closer to the Chilean species have four or five leaflets.

In the axil of each strawberry leaf is an axillary bud that has the potential to develop in any one of three ways, depending on the variety and environmental conditions. One possibility is that the bud may form another crown. If you look at a strawberry plant that is two or three years old, you'll see that it actually looks like a clump of plants growing together, for the original plant has developed additional crowns. Large two-year-old Olympus strawberry plants, for example, may have up to 16 crowns under favorable conditions.

Another option for the axillary bud is to develop into a flower cluster and produce fruit (later we'll discuss the conditions which lead to the important process of flower production). You can already see that the productivity of your strawberry bed depends on how many leaves the plants produce. The more leaves, the more axillary buds; the more buds, the more crowns and flowers. And of course, when your plants have lots of crowns and flowers, that translates into lots of strawberries for you.

It's important for you to know that adequate watering allows your plants to grow unchecked and to produce many leaves. Scientists at Oregon State University in Corvallis found that Olympus strawberries given little water produced 13 percent fewer strawberries than plants that were mulched and generously irrigated. If you're lax in your watering, your strawberries may suffer from water stress, and you'll pay the price the next season when the crop will be far smaller than it should be.

The third possible fate for the axillary bud is that its internodes will lengthen and develop into a runner. If you pay close attention to how your plants are growing, you'll notice that the runners that developed first will produce more flowers the following year than those that came later. That is because they have had more time to grow, and were able to produce more leaves.

FLOWERS

Strawberry flower clusters develop in a most interesting way. To start off, the axillary bud produces two small, leaflike bracts, which are not divided into three parts like leaves. Then, the first, or primary, flower appears. This flower, if properly pollinated, will grow into the biggest berry of the cluster. In the axil of each bract, another bud develops, which grows in the same way as the first bud—it produces two bracts and a secondary flower. This same pattern continues, as each of these bracts, in turn, has a bud that produces two bracts and a third series of flowers. Finally, buds from each of these bract axils may grow to produce one more set of bracts and a fourth group of berries. All in all, as many as 15 berries can be produced in one cluster, but usually fewer than that actually are formed.

When you look at a cluster of strawberries, you'll notice that they're not all the same size. The earliest berries are the largest, and with each succeeding series they become smaller. Scientists at the University of Arkansas studied this difference in size and found that the secondary berries were from 63 to 90 percent as large as the primary one, while the third group of berries were from 35 to 55 percent as big as the primary. When they got to the fourth series they found the berries to be only 25 to 40 percent as large as the primary one. While all strawberry clusters exhibit this

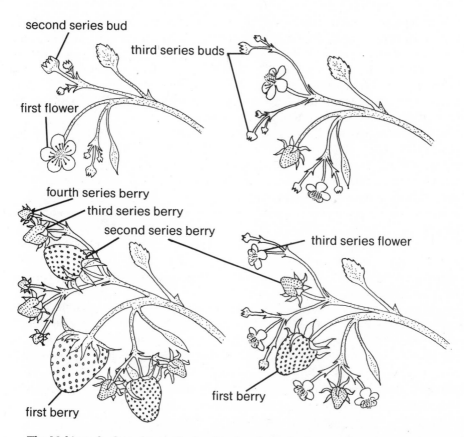

second series bud

third series buds

first flower

fourth series berry

third series berry

second series berry

third series flower

first berry

first berry

The Making of a Strawberry Cluster: Moving clockwise from upper left, you can follow the development of a strawberry cluster. Early in the spring the first flower starts to bloom. As this flower develops into a berry, blossoms for the second series open up and buds for the third series appear. At lower right, as the first and second groups of berries grow larger, flowers for the third series unfold. Some strawberries will even go on to produce a fourth group of berries.

diminishing berry size, the difference is more pronounced in large-fruited varieties. With these, the secondary berries are less than half the size of the first fruit, while the third series of berries may be only 14 percent as big as the primary ones. This size differential is not something that you can control with fertilization or hormone sprays—it's a genetically controlled feature of the plants.

Knowing how berries develop in clusters can ease your disappointment if a late frost kills the primary flowers on your strawberries. The second and third series of berries in the cluster will probably develop just fine, but don't expect them to be as large as the primary berries would have been. If you want to be sure to get those extra-plump first berries, you'll have to be conscientious and cover your plants at night when frost threatens.

RUNNERS

One of the most interesting features of strawberry plants is their ability to send out runners, elongated stems that radiate out from the mother plant to produce a new generation of strawberries. As we mentioned earlier in this section, runners spring from axillary buds. Once runner growth has been triggered, usually two long internodes will be formed before the growing point produces a new crown with short internodes. This new crown then grows leaves that produce roots, and in this fashion, a plantlet is born. If you examine the stem of the runner, you can see the first node, with its tiny, reduced leaves. Normally, this node doesn't grow any farther. But if the runner tip should be destroyed, the axillary bud at that node will spring into action and produce a runner plant.

The elongated runner stem is very well adapted to carrying lots of water and nutrients from the mother plant to the developing plantlet. But this isn't just a one-way street; the enlarged xylem and phloem can actually carry water and nutrients in either direction. This two-way flow is important if a mother plant isn't getting adequate water or nourishment, for its established offspring can send water and nutrients back to it.

Normally, it takes about a month for a new runner to become independent from its mother plant. Until it forms roots, the runner is completely dependent on the parent for water and nutrients, and if the parent plant is water stressed, the runner won't be able to develop roots. This is where you come in. You should always make sure that mother plants setting runners receive ample water. Moist soil encourages the runners to form roots and become independent, so don't overlook the runners, either, when you water.

A LOOK AT HOW BERRIES FORM

Except in the Deep South, strawberry plants break dormancy just about the time the narcissus are in bloom. Soon after that, the first blossoms begin to appear. Cultivated varieties all have perfect flowers, meaning they're self-pollinating. If you examine a strawberry flower closely, you'll see many stamens and pistils. Each pistil contains a single ovule that will develop into a tiny brown seed if fertilized by pollen produced by the stamens.

Although this may all sound normal enough, the strawberry does take an interesting departure from many other fruits in the course of its development. Most fruits we're familiar with have the seed or seeds on the inside. The delicious flesh of fruits like apples and pears develops from the ovary that surrounds the seed as it develops. With strawberries, it's a different story. The juicy red "fruit" grows out of a different part of the flower altogether (actually a modified part of the stem), and the "seeds" themselves form on the outside of the fruit.

What we think of as strawberry seeds aren't actually seeds in the true botanical sense of the word. They're technically fruits called achenes, and each is surrounded by a thin coat that's derived from the pistil. A strawberry may have as many as 400 achenes dotting its rosy surface.

As with other fruits, the developing seeds (in this case, achenes) produce the plant hormone auxin, which causes the fruit to enlarge. So, the more achenes, the bigger the berry. As you might suspect, large-fruited varieties produce more achenes per strawberry than do small-fruited ones. And in individual clusters, the primary berries have more achenes than secondary ones, and so on down the line.

POLLINATION POINTERS

If you're finding misshapen berries or ones that are small nubbins in your strawberry patch, your plants may have a pollination problem. Berries with these defects result from flowers that were incompletely pollinated. In a cool spring, the early blossoms may not be visited often enough by the honeybees and drone flies that are the usual pollinating agents. You can overcome this somewhat by locating your strawberry bed near early blooming fruit trees like cherries or apples or by ringing the bed with early blooming flowers such as grape hyacinths and crocus. These measures should help attract pollinators.

Whatever you can do to lure pollinating insects to your patch will be a help, for without adequate pollination, your strawberry harvest won't amount to much. Experiments in Arkansas demonstrated just how important pollinators are. Researchers found that strawberry plants placed in cages so that insects couldn't pollinate them showed a 40 percent decrease in yield. On top of that, almost half the berries that did develop were malformed.

Besides slowing down the important activity of pollinators, the weather can affect pollination in other ways. Many strawberry varieties won't form mature stamens during the cool weather of early spring. This doesn't mean that they aren't capable of producing berries, though. Fortunately, the flowers remain receptive to pollination for as long as ten days in cool weather. Bees can collect pollen from the third series of flowers and transfer it to the primary flowers whose stamens never developed properly and so complete pollination.

Chilly temperatures wreak havoc in other ways, as well. Temperatures below 60°F may slow pollen germination and growth enough to keep flowers from being completely fertilized. If the mercury plummets even lower, to below 30°F, the pistils can be damaged. In both of these cases, the result is misshapen berries. So, if you live in an area where spring tends to be cool with a good chance of late frost, don't succumb to the temptation to plant the earliest strawberry varieties around, or you may be very disappointed in your harvest.

Of course, nothing is foolproof—even when you plant later varieties, an unusual season can frustrate your desire to eat big, juicy berries. Both of us can attest to this, since we'll be eating mostly the smaller secondary berries this year because

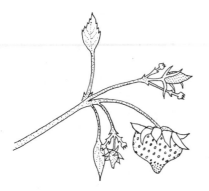

Casualties of Frost: When a late frost visits your garden, it can leave behind some odd-looking strawberries. These are safe to eat but they won't be very flavorful. Even if you're not going to use them, be sure to remove all these irregular berries from the plants.

an early spell of warm weather, which stimulated the strawberries to bloom, was followed by frosts that damaged the pistils. After the frost hit, our gardens contained many strawberry flowers with black centers, which we knew would bear no fruit.

THE FINAL RIPENING STAGE

Assuming that the weather has cooperated and the flower has been successfully pollinated, the next step is for the petals to fall off and the pistils to dry up. After that, you'll have a wait of 25 to 31 days before you can begin to enjoy your harvest. When the weather is warm and days are long, your berries will ripen faster than when days are short or cool.

If you have warm rains around the time your strawberries are ripening, you may be disappointed in the quality of the harvest. The berries are likely to be puffy looking, and when you bite into them they'll be mushy and almost flavorless. Here again, your choice of variety is important. Sparkle is a very popular variety in the North and produces fine, firm berries there. In warm, humid states such as Maryland, however, Sparkle is not even worth growing, for it becomes too soft. In general, no matter what the variety, berries that ripen during a wet, cloudy season just won't have the same sweet burst of intense flavor as berries ripened in sunshine.

PLANNING YOUR STRAWBERRY BED

Don't be in a hurry to order your strawberry plants. Before you rush out and buy anything, you need to sit down and decide just what sort of planting and renewal system you want to use for your bed. Once you've settled that, you'll be able to determine just how many plants you'll actually need.

There seem to be almost as many methods for planting strawberries as there are gardeners! After you decide on one, however, you can't sit back and let your strawberry patch continue for years as originally planted and expect it to keep giving you abundant harvests.

To understand why, you need to know a little bit about strawberry biology. While the original plants you obtain from the nursery will grow vigorously for a couple of years, they usually give out by the third year and produce very few, small berries. This is because their crowns keep dividing and eventually crowd one another out. Meanwhile, they generate a tangle of runners (the number of runners depending on the variety and on conditions), and by the third year, the runners will have made runners themselves.

It's easy to see that without a plan, your strawberry bed will become a crowded, chaotic mess. Beyond the second or third year it won't produce very many berries, and the ones it does produce will be on the small side. You may think that by letting the plants multiply you're bound to get a bigger harvest, but that just isn't so. Not only will the plants be crowded, but they'll have depleted the nutrients in the soil. It's difficult to fertilize a dense strawberry bed since you can't spread compost or manure between the plants. About all you can do is use a liquid spray made from kelp or fish emulsion.

Dorothy learned this lesson the hard way one year when she let a strawberry planting go into its fourth year without thinning the runners or fertilizing the berries. Her Sequoia and Quinalt plants, which should have provided a generous crop of large berries, came through with only a puny harvest of undersized berries. In addition, her already disappointing harvest ended early because a fungus disease spread quickly through the crowded bed and turned the berries to brown mush as fast as they ripened.

Strawberry lovers in the extreme South are put in the position of having to renew their beds *every* year because their mild climate and the dormancy requirements of strawberries aren't compatible. Strawberries need a cold period to break their dormancy, and winter temperatures in the Deep South just don't go low enough to do the trick. There are two ways around this dilemma. You can replant your bed each year with northern-grown plants that have already been exposed to a cold period at the nursery. Or, if you're ambitious, you can perpetuate your own stock by digging up runners and storing them in ventilated bags in the refrigerator for a month before planting them out again.

MATTED ROW SYSTEM

Most gardening books recommend that you use the matted row system, starting a new strawberry bed every two or three years. In some parts of the country, that's a good recommendation for another reason—by the third year, your bed may be contaminated by one of any number of viruses, which will sap the strength of the plants and reduce your yield. If gardeners in your area complain of virus or other chronic disease problems, plan on taking the preventive measure of planting a new bed every second or third year.

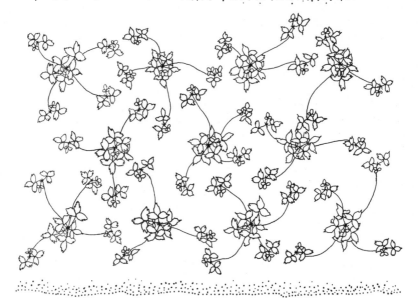

Matted Row System: In areas where virus diseases are a chronic problem, this is the system to use. Allow runners to develop freely and start a fresh bed every two or three years with disease-free stock on a new site.

To get your matted row system started, place your plants 12 to 18 inches apart in all directions. (Use the smaller distance for everbearers, which produce fewer runners, and the larger for the Junebearers like Sunrise and Catskill, which are especially vigorous runner makers.) Leave a 12- to 18-inch buffer zone on either side of the bed for the runners to populate. Don't make the mistake of running more than three rows next to one another, or you'll find that you won't be able to reach over and pick your berries come harvesttime. Stagger the rows so that each plant is opposite an empty space in the adjacent row. The year after making your initial planting, the parent plants and runners will give you an abundant crop of berries. That year or the following one, you should start a new bed with virus-free nursery stock in another part of the garden so you'll have a continuous supply of strawberries in the next few years.

By the third year, your original bed will have become much more crowded because of all those second-generation runners. Depending on the amount of garden space you have and the conditions in your garden, you can either dig up the old bed in the spring, fertilize the ground, and plant a different crop there, or you can delay a bit and harvest the berries before destroying the bed. You will learn from your own experience whether or not you can coax the extra year of production from your plants. If you live in an area where disease problems are minimal, you can renew your patch by using runners from the original plants.

HILL SYSTEM

The merit of this system is that it's a space miser—good news for gardeners short on growing room. The bad news is that it does take time to tend, and it uses more plants than other systems. If you've got more time than growing space, this is probably the system for you.

Start out by planting your berries 12 inches apart in three rows spaced 12 inches apart. Stagger the rows as described for the matted row system. Be sure to leave paths on either side for picking. When runners begin to form, you must be ruthless and remove all of them. This will allow the plants to channel all their energy into becoming established and getting down to the business of producing berries. Your ruthlessness will be rewarded, for these plants will grow quite large, and will bear more berries per plant than the other methods. Depending on your garden conditions, a hill-planted bed will produce well for anywhere from three to six years. As soon as you notice production starting to fall off, order new plants and establish a new bed in a different location.

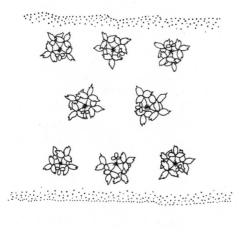

Hill System: This is a space saver, but it does take time to maintain it in good shape. Carefully space plants and leave pathways for picking between every three rows. Remove all runners as they appear.

SPACED RUNNER SYSTEM

The easiest way to keep a strawberry bed healthy is to plant your berries initially in rows 4 feet apart, leaving an extra 2-foot-wide strip of cultivated soil along each edge of the bed. Space the plants 8 to 12 inches apart in the row. During the first year, the runners from the parent plants will fill in the space for a foot or so on either side of the row. Because you allowed such generous spacing between rows

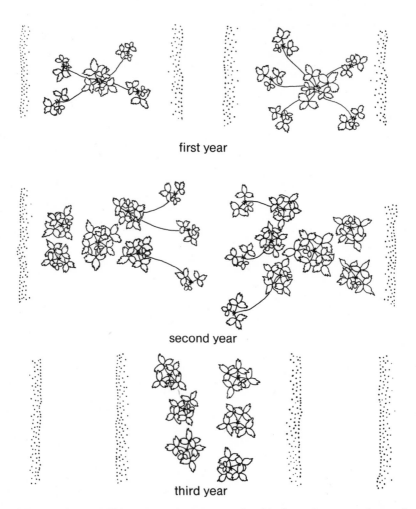

first year

second year

third year

Spaced Runner System: This system takes up room, but it's the easiest way to keep a bed healthy and thriving. Start by planting in rows 4 feet apart. Leave 2-foot-wide paths on each side of the bed. As runners appear, carefully space them 8 inches apart. The second year, let the runners creep into the paths. By the third year, the paths will be full of berry-laden plants. After harvesting carefully, till under the original beds. Leave the plants in the former paths to grow. Every three years, the pathways and growing beds will trade places as you repeat this cycle.

to begin with, you'll still end up with 2-foot-wide paths between the rows of plants. As the runners appear, space them out to 7 to 8 inches apart by pinning the stems to the ground using hairpins, long staples or wooden clothespins. If the plants begin to get crowded, cut off excess runners as needed to maintain the 7- to 8-inch spacing.

The second year, you will have the paths to walk through as you pick the berries. Then, allow the new runners to creep into the paths and the outer edges, but again, space them out. The third year, you can still harvest from all the plants; you'll just have to tiptoe your way very carefully through the bed as you work, since the paths will now be filled with plants. After harvest, till under or dig up the original rows, and treat the areas that started out as paths as your new rows. Spread those dug-up areas immediately with a layer of manure or compost and work it in so the soil will be good and ready the next time you plant in it. From then on, you repeat the harvesting, digging, fertilizing and runner positioning every year or two, depending on how quickly your beds become crowded.

The spaced runner system is probably the most foolproof way of growing strawberries in a perennial bed. It only requires a minimum of special attention when you're setting the runners, and fertilizing the bed is built into the annual routine. Its one drawback is that it does take up quite a bit of room. If your garden is cramped for space, or if you have voracious birds and want to be able to cover your entire bed easily, you'd probably be better off using one of the other two systems.

PLANTING YOUR STRAWBERRIES

Now that you've decided which planting method to use, you need to choose the location for your strawberry bed. Because strawberry plants begin to grow and even blossom quite early, you must be careful not to plant them in a low-lying frost pocket, where a late frost might settle and damage the blossoms. But you don't want to put them in a place that warms up too early in the spring, either, or you'll just be encouraging premature, frost-susceptible growth. A happy medium in areas prone to late frosts is a sunny, north-facing slope. Where frost is not a problem, a south-facing slope will get your plants off to an early start.

Proper drainage is one of the most important considerations in locating the bed. Soggy, poorly draining soil is the kiss of death for strawberry plants. Ironically, spring and fall, the times when these plants are putting on a growth spurt, are also the two seasons when drainage problems are most likely to show up. Avoid trouble right from the start and steer clear of low-lying places where water settles or where you know drainage is poor. If you're caught in the unfortunate position of having nothing but hard clay soil in your yard, start improving the soil for your bed a year ahead of time. Till in lots of compost and some coarse builder's sand to get the soil in tip-top shape for strawberries.

If you have to dig up an area of grass to make room for your strawberry bed, do so the fall before planting. This will expose beetle grubs and wire worms that live among grass roots to winter's cold and remove a possible source of damage to your new plants. If verticillium wilt is a problem in your area, don't plant your strawberries in the same spot where other potential wilt-carrying plants have grown. (For the names of susceptible crops, see Problems with Fruiting Plants in chapter 1.)

PLANTING POINTERS

While strawberries can be planted in the fall in mild climates, fall planting doesn't seem to offer any real advantage except in the South and in parts of California. All in all, it's best to put the plants in as early in the spring as possible. If you've ordered your strawberry plants from a mail-order nursery, they'll probably arrive with bare roots. Try to prepare the bed before the plants arrive so you'll be ready for them. Till in about 5 pounds of barnyard manure for every 10 square feet. You can also use poultry manure, but in smaller quantities—about 1½ pounds per square yard is fine. It's safe to use fresh manure when preparing the bed a full season before planting; otherwise stick to the rotted or well-composted kind so there's no chance of burning the plants' roots.

If you can't have the bed ready and waiting, or if the weather is too wet for planting, see What to Do When Your Plants Arrive in chapter 1 for tips on how to hold your strawberries over in good shape until you can plant them.

A few hours before you begin to plant, soak the roots of your plants in water. Take the soaking container with you out into the garden, and lift each plant from the water just before you place it gently in the ground. This will keep the fine root hairs from drying out and dying. Try to plant on a cool, cloudy day or at least late in the day to give the plants a chance to settle in a bit before having to deal with the drying sun.

The Dos and Don'ts of Planting: If you plant your strawberries as shown at the left and right, you're doing it the *wrong* way. Plants set too low will rot and those set too high are likely to dry out and die. The plant in the middle is the model to follow. Keep the top two-thirds of the crown above the soilline and make sure all roots are covered.

You can doom your plants to an early demise or get them off to a flourishing start—it all depends on the depth at which you plant them. Positioning the plants in the soil is guided by two very simple rules: Keep roughly the upper two-thirds of the crown above the soilline and keep *all* the roots in the soil. When you plant your berries, use a trowel to make a slit in the soil 5 to 7 inches wide and as deep as the roots are long. Fan out the roots as you place the plant in the hole, and gently replace the soil around the roots. After you've planted and watered the whole bed, double-check to make sure the tops of the crowns are still slightly above ground level. This second look is important because sometimes water will wash soil out from around the crown or deposit extra soil there. If necessary, readjust the crown height either by gently scooping soil away from the crown or by adding soil around the base to cover the roots.

Once your plants are in the ground, your job isn't over. Be sure during the first two weeks or so that your bed receives adequate water. It will take some time for the roots to establish themselves, and they need nice moist soil. If the plants look droopy after a day or two in the sun, don't assume the worst—just be patient. Keep the plants well watered, and hope for the best. Dorothy had a scare one year when she set out her new plants on an unusually sunny spring day. She feared that half of them had died, for the leaves had dried up and turned brown. But in a couple more days, new leaves appeared, and her plants were off to a good start after all.

CARE THROUGHOUT THE SEASON

Most garden experts recommend that you remove the blossoms that show up on your strawberry plants the first season. For Junebearers, this means no crop at all the first year. With everbearers, you can pluck just the first set of flowers and leave the subsequent blooms destined to give you a fall crop. The plants will be far enough along by then to bear without it being a severe strain on their resources.

Plants that are allowed to blossom freely the first year don't form feeder rootlets as quickly as plants that have been deflowered. This can affect how well the plants become established; the better the root system, the better the nutrition the plant receives. Once the blossoms are removed, the plants don't need to invest any energy into fruit formation and can become bigger and stronger—primed for producing berries the next season. What you're actually doing is delaying gratification—your plants will end up producing far more berries the second year than they would have if allowed to set fruit the first year.

Throughout the season, be sure to keep your strawberries watered. Since the roots occupy the top 6 inches of soil, they're very vulnerable to drying out in a hot, arid climate or in sandy soil. Water-stressed strawberry plants are dull in color and don't set fruit well—they often produce deformed and small berries. A steady supply of water is important as the fruits are developing, too, for they contain a high percentage of water. And while the plants are forming flowers for the next year's crop, lack of water will result in a decrease in the number of blossoms. Runners have trouble rooting in dry soil, but establish themselves quickly in moist soil. All this should be enough to convince you to keep the top 2 inches of soil evenly moist from the time you plant your berries all the way through the growing season.

Because they depend on a steady water supply all season long, strawberries are a prime crop for drip irrigation. Overhead watering, especially if you're in a humid area, can do more harm than good. With this method, you risk wetting the berries, which makes them puffy and softens the skin. Berries in that condition are susceptible to fungus growth, and usually rot. The slow, steady delivery of water through a drip irrigation system prevents all these problems. If it's not feasible to install a system and the best you can do is to water with a sprinkler, water during the warm part of the day so the plants dry out quickly.

After the soil warms up, apply a mulch like straw around the plants just following a heavy soaking. Don't go overboard with the mulch and cover the crowns; that's an open invitation for disease problems. If you want to discourage runners from setting, consider using a black plastic mulch.

Strawberries need rich soil to produce abundant crops. Each time you renew the bed, you should work in manure at the rate we recommend in the preceding section on planting. Plants that are getting just the right amount of nitrogen will have lots of large leaves, which in turn will enable them to make more crowns and flowers. Too little nitrogen will show up as yellowing leaves and slow growth. You can give them too much of a good thing, though; excess nitrogen will actually trigger more vegetative growth at the expense of berries, and the plants won't be as hardy. Strawberries also need potassium and phosphorus; usually, between the mulch and the manure they get just what they need. If they don't, they'll let you know by producing few flowers and by not being very winter hardy. If you suspect a deficiency, turn to chapter 1, The Nutrients Fruiting Plants Need, for some good sources of these nutrients.

When fall rolls around, you should give some thought to winter care for your plants. If you live in an area where winters are harsh, start out with super-hardy varieties such as Dunlap, Fort Laramie and Ogallala for extra insurance that the plants will make it through. Both Dunlap and Ogallala can survive temperatures of −40°F without any protection. But in an area where harsh winters don't always bring a snow cover, just bitter temperatures, or where there are frequent thaws followed by sudden drops in temperature, you need to give your plants the added protection of a hay or straw mulch. Both of us have gotten by many years without protecting our berry plants—but that's only because in Missoula there's almost always a snow cover during the most severe part of the winter.

Cover your plants only after the ground has frozen hard. Make sure the mulch layer is 3 to 4 inches thick. As soon as spring temperatures start to warm the ground, it's time to remove the mulch—but keep it handy. You may need to throw it back over the plants if they start to bloom and the temperature plummets back down for a day or two.

STRAWBERRIES FROM SEED **55**

STRAWBERRIES FROM SEED

The latest thing in strawberries is Sweetheart, a berry you can start from seed in the dead of winter and transplant to your garden in spring. The plants will produce sweet, juicy berries about 140 days after you sow the seeds. People who've grown these plants claim the berries are the sweetest they've ever eaten—they're so good tasting that even the birds show an extra interest in them!

To get them started, scatter the tiny seeds over the top of a flat filled with potting soil. Sprinkle a very thin layer of soil over them, water with a fine mist, and cover the flat with a sheet of plastic to hold in moisture. (Since the seeds are close to the top of the soil, they can dry out very easily, so be careful.) Place the flat in a cool place—about 65°F—and keep the soil evenly moist. If the temperature gets above 75°F, the seeds won't germinate well.

It takes a long time for all the seeds to germinate, so you must be patient. After 30 days, most of them will have sprouted. Remove the plastic as soon as you see the sprouts and place the flat in bright light. The best temperatures are from 60° to 70°F during the day and from 55° to 60°F at night. Once the seedlings have from three to four leaves, carefully thin or transplant them so that they are about 6 inches apart, or put them into small, individual pots. Keep them well watered, and feed them with a liquid fertilizer such as manure tea or fish emulsion once a week. As springtime arrives, harden off the seedlings by placing them out in the sunshine. Start out with just a few minutes a day and work up to a full day out-of-doors. When the chance of frost is past, transplant your strawberries to the garden.

If you plant Sweetheart seeds in February, your strawberries should begin producing flowers and runners in June. Instead of removing the flowers, cut off all the runners. That way, the plants can put their energy into producing fruit their first year. From this point on, you can treat Sweetheart just like any other ever-bearer. Sweetheart appears to be widely adapted and gardeners from the far North to the Deep South have been able to grow it. It is not sensitive to daylength, which makes it capable of producing fruit in all areas.

Sweetheart isn't the only strawberry you can start from seed. For several years, seeds of the *fraises des bois* or alpine strawberry have also been available. Alpine strawberries are decorative plants that reproduce by crown division rather than by runnering. Their fruits are small but especially flavorful, and the plants produce berries all summer long. Alpine strawberries make great edging plants for garden walkways, providing a hungry gardener with a luscious snack between chores.

You can start alpine strawberries from seed as we described above, or you can purchase plants from a nursery. Whether you grow your own from seed or buy plants, the culture is the same. Space the seedlings a foot apart in all directions and water them well. They should receive about an inch of water a week. You'll need to divide the plants every two years. In the spring, just as the crowns are showing their first flush of new growth, dig them up, separate them into three to five pieces (distributing the roots evenly), and replant them immediately. Work with one crown at a time so that the roots don't dry out. Alpine strawberries are cold hardy, but you should mulch them during the winter to protect them from alternate freezing and thawing.

HARVESTING AND STORING THE BERRIES

When the long-awaited time has come and your berries have lost all traces of green and are plump and red and ready to be harvested, pick them carefully. Bruised berries will lose more vitamin C than undamaged ones, and they'll also spoil much faster. Unless you're going to eat the fruit right away, leave the little green caps on. Removing the caps is like pulling the flavor plug on your strawberries! You also affect their nutritional value. Berries without caps can lose 90 percent of their vitamin C at room temperature in only two days. With the caps gone, fungus and bacteria have a prime site at which to invade, spoiling your berries before you get a chance to enjoy them.

Make sure you pick your berries as soon as they are ripe, otherwise the immature ones won't ripen as fast. Pick the damaged and overripe berries too, or they will rot on the plant. Once there are even a few rotting berries in the patch it's all too easy for fungus to spread and strike down perfectly healthy berries. Diane found out just how important this is one year when she went away on vacation and left a neighborhood child in charge of picking. When Diane returned, her strawberry patch was almost ruined, for it was full of brown, slimy, rotten berries. Apparently, the berries didn't get picked as often as they should have been, and the mature ones started to rot. Diane and her husband had to spend hours cleaning out the unpleasant mess, and the next year the bed was plagued with brown rot—something they had never had problems with before. Since then they've relocated the entire bed and haven't had any difficulties—especially since they're careful to pick *all* the berries as they ripen.

Once you've picked them, eat the berries as soon as possible. The tender, juicy varieties developed for home gardens don't keep as well as the hard, dryish ones grown commercially. If you must store the berries, stash them unwashed in the refrigerator in a covered container. Wash them right before using, and wait to remove the caps until after they've been rinsed to avoid making the berries soggy.

PROBLEMS WITH STRAWBERRIES

With any kind of perennial strawberry bed, you must always remain on the alert for disease problems. If you lose a significant part of your crop to fungus, you should replant in another area and be especially careful to avoid overcrowding in the new bed. If your yield decreases as the years go by despite generous fertilization, you might suspect a resident virus and replant in a different area with fresh, virus-free nursery stock. This same advice applies as well to any other disease problems that should appear on the scene. The best preventive measure you can take is to start out with plants that offer some disease resistance. We mention which varieties have built-in disease protection in the Guide to Strawberry Varieties at the end of the chapter.

In the strawberry patch, problems are likely to start out underground since many strawberry diseases attack the roots. Because the roots play an important role in gathering and storing nutrients, strawberries afflicted with root diseases will grow weak and produce poorly. If you see that your plants are growing slowly, even though you have fertilized them and provided plenty of water and sunlight, you should suspect a root disease. Here's a rundown of some of the more common ones.

Red Stele: This is one of the most devastating strawberry afflictions. It stunts the plants and sometimes kills them just before the berries ripen, a great frustration for the expectant gardener. If this happens to your plants, you can do a postmortem examination by checking the color of the root center. Slice the root crosswise with a sharp knife or razor blade. A center that is brownish red instead of the normally healthy yellowish white will reveal red stele as the culprit.

Once you've made the diagnosis you should remove any plants you suspect are infected and discard them; don't add them to the compost heap. There's no treatment yet to control this disease, so if it's a problem in your strawberry patch, try a resistant variety next time.

Black Root-Rot: When a cross-section of root reveals a blackish or dark brown center, or the roots have dark spots, suspect black root-rot. The exact cause of this disease is unknown, but it's associated with soil fungi, nematodes, winter injury, fertilizer burn, drought or too much salt or water in the soil. Since there's no way to control the disease, again, all you can do is dig up the sick plants and start over with healthy ones. If you suspect any of the associated conditions exist in your garden, do what you can to change them.

Verticillium Wilt: Strawberries infected by this fungus show symptoms in their leaves; they wilt and dry at the margins and between the veins. Their growth is stunted and runners can develop black streaks. The best way to avoid this problem is to use resistant varieties. For more on verticillium wilt, see Problems with Fruiting Plants in chapter 1.

Strawberry foliage is also susceptible to disease, with leaf scorch and leaf spot being the most prevalent. Leaf scorch shows up as small, dark purple spots, while leaf spot is recognizable as purple spots with white or gray centers. If either of these diseases is a problem in your area, look for varieties that are resistant.

One pest to be on the lookout for is the strawberry crown borer. It does the most damage in its grub form, by boring into the main roots. Since grubs are hatching from March through August, the damage can go on all summer long. Be on guard against any small, brown beetles you see on your plants—they're the adults laying eggs. If your bed becomes infested, plant a new one at least 300 yards away. Since the beetles don't fly, they won't come along. Also, keep wild strawberry plants and cinquefoil away from the bed—these are also host plants for this pest.

STRAWBERRY FRONTIERS

Researchers are hard at work developing new strawberry varieties, but unfortunately this isn't completely good news for home growers. Much of their labor is concentrated on producing strawberries that ship well for the commercial growers in California, rather than on better varieties for backyard gardeners. However, more disease-resistant varieties should appear in our nursery catalogs in the future as a result of this breeding for commercial growers.

One interesting development in strawberry breeding is the increasing number of varieties, such as the new Tristar and Tribute (see the Guide to Strawberry Varieties), which are not only very disease resistant but are also daylength neutral. This means they don't respond to daylength in the ways conventional varieties do and can continue to produce over a long period of time. These varieties are generally listed under everbearers in catalogs, and they're easy to recognize by their descriptions as plants which produce over a longer season than other varieties.

Guide to Strawberry Varieties

EVERBEARING VARIETIES

Fort Laramie—Medium to large, firm, midseason. Resists leaf spot and leaf scorch; susceptible to red stele and verticillium wilt. Grows best in Central states, North and West. Good fresh, freezes well.

Gem (Superfection)—Small, soft, midseason. Resists leaf scorch; susceptible to leaf spot and red stele. Fair quality fresh or frozen.

Ogallala—Medium size, medium-firm, midseason. Susceptible to red stele. Hardy, grows best in North. Very good fresh, freezes well.

Ozark Beauty—Medium size, medium-firm, late. Resists leaf spot and leaf scorch; moderately resistant to verticillium wilt; susceptible to red stele. Grows best in East, North, Midwest. Very good fresh, freezes well.

Quinault—Medium, soft, midseason. Resists leaf spot, leaf scorch, red stele; susceptible to virus. Grows best in California, North, Northwest. Good fresh, fair when frozen.

Tribute—Medium to large, very firm, midseason. Resists red stele and verticillium wilt; tolerant of leaf spot and leaf scorch. Grows in East, Midwest, Northwest, South. Very good fresh, freezes well.

Tristar—Medium, firm, early. Resists red stele and verticillium wilt; tolerant of leaf spot and leaf scorch. Grows in East, Midwest, Northwest, South. Very good fresh, freezes well.

JUNEBEARING VARIETIES

Albritton—Large, very firm, late. Very resistant to leaf scorch; resists leaf spot; susceptible to red stele, verticillium wilt, virus. Grows best in South. Excellent fresh, freezes well.

Allstar—Extra large, firm, midseason. Resists leaf spot, leaf scorch, red stele, verticillium wilt; tolerant of virus. Grows in East, Midwest, Southeast. Very good fresh and frozen.

Apollo—Large, very firm, midseason. Very resistant to leaf scorch; resists leaf spot; susceptible to red stele and verticillium wilt. Grows best in South. Good fresh or frozen.

Atlas—Extra large, firm, early. Very resistant to leaf scorch; resists leaf spot; moderately resistant to verticillium wilt; susceptible to red stele. Grows best in Southeast. Good fresh, freezes poorly.

Blakemore—Small, firm, early. Tolerant of virus; susceptible to leaf spot, red stele, verticillium wilt; very susceptible to leaf scorch. Grows best in South. Fair quality when fresh, freezes well.

Catskill—Extra large, soft, midseason. Very resistant to verticillium wilt; resists leaf scorch; susceptible to leaf spot and red stele; very susceptible to virus. Grows best in Midwest and Northeast. Good fresh, fair when frozen.

Cyclone—Large, soft, early. Resists leaf spot; tolerant of virus; susceptible to red stele. Grows best in Midwest. Very good fresh, freezes well.

Dabreak—Medium size, medium firm, very early. Very resistant to leaf spot; resists leaf scorch; tolerant of virus; susceptible to red stele and verticillium wilt. Grows best in South. Good fresh or frozen.

Darrow—Medium, firm, early. Resists red stele; moderately resistant to verticillium wilt; susceptible to leaf spot and leaf scorch. Grows best in East. Very good fresh, freezes well.

Delite—Large, firm, late. Resists leaf spot, leaf scorch, red stele, verticillium wilt. Grows best in Southeast and South Central states. Fair quality fresh or frozen.

Earliglow—Medium to large, firm, very early. Resists red stele and verticillium wilt; moderately resistant to leaf spot and leaf scorch. Grows best in Midwest. Very good fresh or frozen.

Fairfax—Medium, firm, midseason. Resists leaf spot and leaf scorch; susceptible to red stele and virus. Grows best in North. Excellent fresh, fair when frozen.

Fletcher—Medium, medium-firm, midseason. Very resistant to leaf scorch; resists leaf spot; susceptible to red stele and verticillium wilt. Grows best in North. Very good fresh, freezes well.

(continued on next page)

Guide to Strawberry Varieties
Continued

JUNEBEARING VARIETIES

Florida Ninety—Extra-large, soft, midseason. Very susceptible to leaf spot and leaf scorch; susceptible to red stele, verticillium wilt. Grows best in South. Very good fresh, fair when frozen.

Guardian—Extra-large, firm, midseason. Very resistant to verticillium wilt; resists leaf spot, leaf scorch, red stele. Grows best in East. Good fresh, fair when frozen.

Honeoye—Extra-large, firm, midseason. Resists leaf spot and leaf scorch. Grows best in Midwest, North. Good fresh, freezes well.

Hood—Large, medium-firm, midseason. Resists leaf spot, leaf scorch, red stele, verticillium wilt; susceptible to virus. Grows best in Northwest. Very good fresh, freezes well.

Midland—Large, firm, very early. Resists leaf spot and leaf scorch; susceptible to red stele, verticillium wilt, virus. Grows best in Midwest, Northeast. Excellent fresh, freezes very well.

Midway—Large, firm, midseason. Resists red stele; moderately resistant to verticillium wilt; susceptible to leaf scorch; very susceptible to leaf spot. Grows best in Northeast, Northwest. Good fresh, freezes very well.

Northwest—Medium, medium-firm, large. Resists leaf spot; moderately resistant to verticillium wilt; tolerates virus; susceptible to red stele. Grows best in Northwest.

Pocohontas—Large, medium-firm, midseason. Resists leaf spot; moderately resistant to leaf scorch; susceptible to red stele and verticillium wilt. Grows best in East, South. Good fresh, freezes very well.

Raritan—Large, firm, midseason. Susceptible to leaf spot, leaf scorch, red stele, verticillium wilt, virus. Grows best in Midwest and Northeast. Fair fresh or frozen.

Redchief—Large, firm, midseason. Resists leaf spot, leaf scorch, red stele; moderately resistant to verticillium wilt. Grows best in Midwest. Good fresh, freezes very well.

Red Star—Large, firm, very late. Resists leaf scorch; moderately resistant to verticillium wilt; tolerates virus; susceptible to leaf spot and red stele. Grows best in East. Good fresh, freezes well.

Robinson—Large, soft, late. Resists verticillium wilt; moderately resistant to leaf spot; susceptible to leaf scorch, red stele. Grows best in Southeast. Fair fresh or frozen.

Scott—Extra-large, firm, midseason. Resists leaf spot, leaf scorch, red stele; moderately resistant to verticillium wilt. Grows best in East, Midwest. Very good fresh, freezes very well.

Sequoia—Extra-large, soft, very early. Tolerates virus; susceptible to red stele, verticillium wilt. Grows best in California, Northwest. Very good fresh.

Shuksan—Large, medium-firm, very late. Resists red stele, verticillium wilt; tolerates virus. Grows best in Northwest. Good fresh, freezes very well.

Sparkle—Small, soft, late. Resists red stele; moderately resistant to leaf scorch; susceptible to leaf spot, verticillium wilt, virus. Grows best in Midwest, Northeast. Very good fresh, freezes very well.

Sunrise—Medium, firm, very early. Resists leaf scorch, red stele; moderately resistant to verticillium wilt; susceptible to leaf spot. Grows best in East, South Central states. Very good fresh, fair when frozen.

Surecrop—Large, firm, midseason. Very resistant to verticillium wilt; resists leaf spot, leaf scorch, red stele; tolerates virus. Grows best in East. Good fresh, freezes well.

Tangi—Medium, medium-firm, midseason. Resists leaf spot and leaf scorch; susceptible to red stele. Grows best in South. Good fresh.

Tennessee Beauty—Small, firm, late. Resists leaf spot and leaf scorch; tolerates virus; susceptible to red stele. Grows best in South. Good fresh, freezes well.

Tioga—Extra-large, firm, midseason. Tolerates virus; susceptible to leaf spot, red stele, verticillium wilt. Grows best in Northwest, Central states. Good fresh, freezes well.

Totem—Large, medium-firm, late. Resists red stele; tolerates virus. Grows best in Northwest. Good fresh, freezes well.

Trumpeter—Medium, soft, midseason. Tolerates virus; very susceptible to leaf spot; susceptible to red stele. Grows best in Midwest. Good fresh, freezes very well.

Tufts—Large, firm, midseason. Moderately resistant to leaf spot; tolerates virus; susceptible to red stele and verticillium wilt. Grows well in California. Good fresh.

CHAPTER 3

BRAMBLE FRUITS
NOT AS THORNY
AS YOU THINK

Vital Statistics

FAMILY

Rosaceae

SPECIES

Rubus idaeus (red and yellow
raspberries)
R. laciniatus (evergreen or cutleaf
blackberry)
R. occidentalis (black raspberry)
R. procerus (Himalaya giant blackberry)

POLLINATION NEEDS

Self-pollinating

HOW LONG TO BEARING

First large crop will be harvested
between 2 and 3 years following
planting.

CLIMATE RANGE

Red and Yellow Raspberries
Zones 3–10
Blackberry (erect)   Zones 4–10
Blackberry (trailing   Zones 5–10
Black Raspberry   Zones 5–8
Purple Raspberry   Zones 4–8

For many people, the word bramble conjures up images of thorny, tangled masses of unruly canes. Wild blackberries certainly fit this description, but the modern, cultivated varieties of bramble fruits—which include raspberries and various sorts of blackberries—have been tamed to suit the home grower's needs. Modern breeding has given these plants a new image; they are easy to contain within neat rows, and there are many thornless types, as well as varieties with fine, soft thorns.

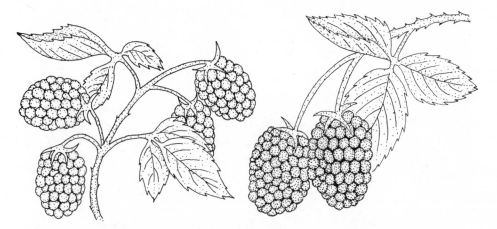

Mouth-Watering Brambles: The hardest thing about growing raspberries (left) and black-berries is the harvest, for these fruits are so delicious that you'll want to eat them on the spot instead of bringing them inside to share with others.

Raspberries and blackberries vie with strawberries as the easiest, most reward-ing home fruits to grow. They're good choices for impatient gardeners because they begin to produce the second year, the same as Junebearing strawberries. Although bramble fruits don't yield as generously at first as strawberries do, brambles have the advantage of providing a permanent planting. As long as no serious disease problems develop, you can keep your rows of blackberries and raspberries in the same place for ten or more years. Brambles are relatively carefree and give you a generous crop of juicy berries for a minimum amount of fuss. They take up rela-tively little space, too—a 25-foot row can supply plenty of berries for a family of four.

Probably the most compelling reason to grow your own bramble fruits is a simple one—that's the only way you'll get to enjoy these ambrosial berries. Black-berries and raspberries hardly ever show up in produce markets, and when they do, the prices are sky-high. The reason these berries command top dollar is that they're so fragile they must be hand-picked, and their shelf life is perilously short. But when you've got a row of bramble fruits in your backyard, you don't have to worry about things like shelf life and outrageous prices—you just have to "worry" about not eating too many at one time.

UNTANGLING THE BRAMBLES

There are many common names for the various brambles, which can be very con-fusing. Not only are the common names mixed up—botanists can't agree on how to

classify the brambles either! About the only thing they agree upon is that all these berries belong to the genus *Rubus.* The main reason for the confusion is that brambles
hybridize easily, and when they do, they often develop very strange genetic
characteristics.

SOME BRAMBLE BACKGROUND

Around 1850, the Europeans brought two blackberry species to America; the evergreen or cutleaf blackberry (*Rubus laciniatus*) and the Himalaya giant blackberry
(*R. procerus*). These plants escaped from cultivation and hybridized with one
another and with the native American blackberry species to create the incredibly
vigorous blackberries that plague the West the way kudzu afflicts the Southeast.
Not only are these prickly, tangled vines hard to control, but half of them produce
no fruit at all, since the male and female vines are separate. Few residents of the
West Coast are interested in growing blackberries in their gardens—since they can
pick them in abundance from the wild, why bother? But they really should consider
making room in their gardens for the more sedate cultivated bramble varieties. The
wild vines are bristling with thorns and depend on natural rainfall to produce
berries, while garden-grown canes can be blessedly thornless and produce a reliable
crop of big berries.

WHAT'S IN A NAME?

Now let's take a look at the common names for brambles and sort out what they
really are, as much as possible.

Red Raspberries: The cultivated red raspberry we know today was developed
from hybrids between two species, *Rubus idaeus,* from northern Europe and Asia,
and *R. strigosus,* from North America. While wild red raspberries require cross-
pollination, the cultivated varieties can pollinate themselves, so you don't need to
plant more than one variety. Red raspberries come in thorny and thornless varieties,
and some produce two crops a year.

Yellow Raspberries: These raspberries were developed from red varieties; their
flavor is sweeter and milder, but their cultural requirements are the same as for
the red type. Like red raspberries, some yellow varieties produce two crops a year.

Black Raspberries: Also called black caps, these raspberries are a cultivated
version of a native American vine, *Rubus occidentalis.* Juice from blackcaps is used
for stamping grade markings on meat and once provided a major flavor component
for the soft drink, Dr. Pepper. Now, the same chemical is extracted from plums
instead, which are cheaper than black caps.

Purple Raspberries: This berry is a hybrid between red and black raspberries. In
growth habit and cultural characteristics, purple raspberries are similar to black ones.

Blackberries: Like the wild-growing western blackberries, cultivated varieties
are derived from hybrids among several wild species. But unlike the wild vines,
cultivated blackberries have flowers with both male and female parts, so all vines
produce fruit.

Blackberries may be either thorny or thornless. The thornless varieties were
developed in two different ways. Some, such as thornless Evergreen, carry the

genetic information for the thornless characteristic only in the outer layer of cells on the cane. When one of these vines dies back to the roots or is injured, new cells for the outer layer of the cane develop from the inner part of the vine. When these replacement cells take over, thorny canes can develop. Other thornless types, such as Smoothstem and Thornfree, don't have this problem because all the cells of the vine carry the thornless trait.

Blackberry canes grow in two basic ways. Some varieties produce upright canes that don't become especially long. Others, called dewberries in the South and better known as western trailing varieties in the West, produce flexible, trailing canes which can grow to 10 feet or more in length. Varieties intermediate between upright canes and trailing vines have also been developed.

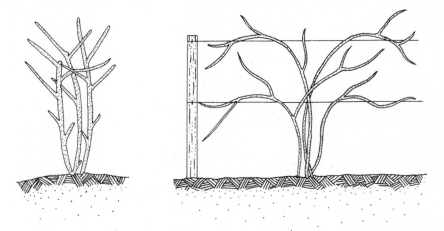

Two Kinds of Blackberries: Blackberry vines come in two very different forms. One kind, shown on the left, produces upright canes that stay on the short side. The other kind, the trailing blackberry, develops long, flexible canes that can sometimes reach 10 feet or more in length.

Loganberries, Youngberries and Boysenberries: All three of these berries are actually just varieties of trailing blackberries. The loganberry was developed in California by J. H. Logan in 1881, while the youngberry originated in North Carolina in 1926. Boysenberries are somewhat of a mystery, since nothing is really known about their origin. As far as flavor and appearance are concerned, these three have some distinguishing characteristics. Loganberries are elongated and reddish purple, with a notable but not unpleasant, acidic taste. Both youngberries and boysenberries are a shiny purple-black. Boysenberries are especially large and have a distinctive aroma.

A UNIQUE TASTE SENSATION

With the ease of hybridization among the various species of *Rubus,* you may wonder what a cross between the rich, almost wine-flavored blackberry and the tangy raspberry would taste like. Your mouth may water at the flavor possibilities of such a union, but one of the basic differences between the fruits of the two vines has made it difficult to produce a desirable hybrid.

Each blackberry or raspberry is the result of pollination of one flower. The fruit is made up of tiny fruitlets, called druplets. Each druplet contains one seed. The cluster of druplets which makes up the berry is anchored to the vine by the fruit receptacle. When the raspberry is ripe, it separates from the receptacle and slides off easily. The blackberry fruit, on the other hand, doesn't part ways with the receptacle. When the berry is ripe, the receptacle lifts easily from the vine, so you pick and eat the whole thing—receptacle as well as druplets.

Over the years, when scientists have hybridized blackberries with raspberries, the resulting fruit has tended to combine the wrong traits of the parents. The druplets stick to the receptacle, as in blackberries, but the receptacle sticks to the vine, as in raspberries. What you've got, then, is a fruit that's very difficult to wrest from the vine. To date, the only successful hybrid is the tayberry, a dark purple fruit that looks like a large loganberry.

BRAMBLE VARIETIES

Hardiness is a major concern and will probably guide your selection of a bramble variety. Red raspberries are the hardiest of the brambles; so if you live in the North, they're a good crop to choose. Latham is the most widely grown variety and is quite hardy. Canby Thornless is a favorite of ours. It produces abundant, flavorful berries on thornless, vigorous, hardy canes.

The two-crop varieties of red and yellow raspberries are very satisfying to grow, for they give you a double harvest each year. However, in short-season areas, these varieties may not produce much in the way of a second crop. Fall Gold raspberries are usually just beginning to produce heavily in our gardens when frost catches them in mid-September. August Red ought to beat the frost in the North, since it begins to produce earlier than other two-crop varieties. For gardeners who don't have to worry so much about early fall frosts, Heritage is an especially high-quality fall bearer which tends to produce somewhat later than most.

Black and purple raspberries aren't as hardy as red and yellow types, but some varieties will grow quite well in the North. The hardiest varieties of upright black-berries are often difficult to locate, but a few available varieties will produce well in cold climates, such as Darrow and Ebony King. Another problem with black-berries is that some are so late to produce that the berries are nipped by frost before they ripen.

Northern gardeners should take note that the trailing types of blackberries are less hardy than the upright ones. Dorothy, unfortunately, found that out only after she planted some in her yard. For several years, she has managed to keep a few hills of thornless boysenberries going. The vines grow vigorously during the summer, and she carefully covers them with leaves every autumn. But come springtime, the agonizing decision about when to uncover them and expose the tender buds to late cold weather must be made. No matter when Dorothy gambles that the worst of the cold is past, she ends up losing half the vines. Some are killed outright by the cold, but an equal number are damaged in the process of uncovering and manipulating the brittle, trailing vines. The fact that Dorothy is stubborn is the only thing that keeps her from ripping up those plants altogether and replacing them with hardier ones. If she ever does, she'll replant with hardier upright blackberries.

In the South, it's not the climate that poses problems for western trailing types. These blackberries won't survive there because they're susceptible to endemic diseases. However, the trailing dewberries are able to tolerate those diseases better.

As you can see, you really can't just rush out and buy any type of bramble that strikes your fancy. The region you live in does present limitations. For suggestions about which varieties of brambles to grow in your area, see the Guide to Bramble Varieties at the end of the chapter.

HOW BRAMBLES GROW

No matter what you call them, all brambles share a common growth pattern. In the spring, new canes grow from basal buds on the underground portions of year-old canes. Except in the fall-bearing raspberries, these new canes, which are called primocanes, don't bear fruit. The primocanes of red and yellow raspberries and thornless blackberries have few, if any, branches. But in black and purple raspberries and some blackberries, the primocanes do branch. Since the branches will carry the blossoms the next year, you should top the canes of these branching plants the first year they appear. By doing this, you're setting the stage for a bumper crop of blossoms the following year. (See Fearless Pruning later in this chapter for pointers.)

The short days and cool temperatures of fall stimulate the flower buds for the next year's crop to form. As with other fruits, the flower buds reach a certain stage and won't continue developing until they've gone through a cold dormant period. Northern varieties require a temperature of 45°F or lower to go into dormancy, while southern varieties will go dormant at higher temperatures. Trailing blackberries grow well in the South and coastal West precisely because they don't require a very low temperature to undergo dormancy. In the North, these fruits have difficulty surviving, for their chilling requirement may be met quite early in the winter. Then, if a warm spell comes along, they'll be ready to come out of dormancy. Once they do, they're susceptible to winter freeze damage if the temperature should drop again. In general, red raspberries require more winter chilling than do black raspberries and blackberries. This makes them hardier in the North.

The second year, when the brambles awaken from their winter sleep, they develop new primocanes. The second-year canes, which were primocanes the preceding year, are now called floricanes, for they will flower and produce berries. After the berries have ripened, the floricanes decline and die.

The exceptions to this general course of events are fall-bearing raspberries such as August Red, Fall Gold and Fall Red. These plants produce berries late in the season on the tips of the primocanes. When a primocane reaches a certain height,

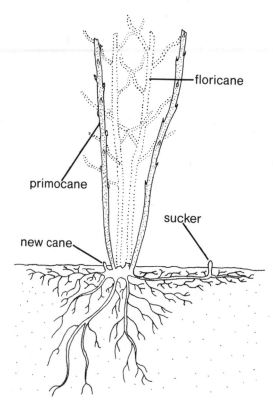

floricane

primocane

new cane

sucker

How a Bramble Grows: This raspberry plant can give you a general idea of how brambles grow. In early spring before the plant leafs out, you can see the primocanes clearly. These year-old canes will produce berries later in the season. The dotted lines shown the floricanes you pruned away last summer. Those are the canes that gave you a harvest that season. On the left, a new cane is peeking through the soil. New canes form from buds on underground portions of the primocanes. On the right, a sucker is emerging from the roots.

the entire process of flower development (from blossom to berry) occurs at its tip. Meanwhile, the normal process of flower bud development is going on in the rest of the cane. In the springtime, after undergoing a dormant period, the lower portion of the cane (now, technically a floricane) will burst into bloom, like other brambles, while the tip remains bare. Commercial growers sometimes take advantage of this characteristic, turning fall-bearing raspberries into an annual crop. They plant in the spring, harvest in the fall, and plow the vines under after harvesting. This allows the growers to avoid the pest and disease problems that can plague large, permanent raspberry plantings.

A LOOK AT FACTORS THAT AFFECT YIELD

Before we go into the specifics of planting and caring for bramble fruits, you should know something about how growing conditions can affect yield. Scottish scientists have recently found that the most important factors in red raspberry production are spacing and access to light. They discovered that the optimum density of canes was eight to nine for every 3 feet in length of a 2-foot-wide row. When each cane received its full share of sunlight, it was able to manufacture enough food to produce a high yield of berries. However, if the canes shaded one another, then each cane produced fewer berries.

We have noticed an interesting phenomenon relating to this in our own gardens. Both of us grow Canby Thornless obtained from the same source. In Dorothy's garden, where the canes are spread out in one long, narrow row, berries are borne all the way down to near the bottoms of the canes. But in Diane's garden, where the plants grow in several close, short rows that partially shade one another, raspberries are only produced on the upper halves of the canes.

Ample light is also important for boysenberries. Commercial growers in Oregon generally bundle boysenberry canes together before placing them on a trellis, while California growers spread out the canes. Although the growing season in California is warmer and longer than in Oregon, it is hard to believe that climate alone can account for the fact that California boysenberry plantings produce almost two-and-a-half times as much as Oregon plantings do. Chances are that the bundled Oregon canes shade one another considerably, while the spread-out California plants receive much more light, resulting in more berries.

Bramble canes are capable of producing more flowers than they actually do. Factors such as the nutritional status of the plants, the length of the cane after pruning and the degree of winterkill all affect the number of flowers that blossom and the number of berries that set. For example, thick canes produce more berries than thin ones, basically because they contain more stored carbohydrates. When flowering begins in the spring, the plants are still relying in part on stored carbohydrates for nourishment. The thinner canes have less to offer the developing flowers, resulting in a lower degree of berry set.

When you prune, you're affecting yield, but not quite to the degree you might expect. It's true that pruning cuts down on the potential number of berries you could be enjoying. But the plants can compensate somewhat for the lost potential flowers by allowing a few more blossoms to emerge on the remaining portion of the cane.

CHOOSING A SITE AND SPACING BRAMBLES

Because your bramble planting will remain in the same location for many years, take plenty of time to consider where the best spot for it is. Try not to plant brambles where another crop susceptible to verticillium wilt has grown (see Problems with Fruiting Plants in chapter 1 for a rundown of these crops). If that's not possible, at least try to wait three years before planting brambles on the site. And here's another consideration: Keep brambles at least 300 feet away from any wild raspberries or blackberries. Wild plants are very likely to harbor pests and diseases that could be passed on to your garden crops.

Before you can figure out how many plants to order, you must decide how to space your plants and how you'll train them. Both spacing and training vary considerably, depending on which bramble crop you're planting. Generally, brambles are planted either in hills or rows. (There's more on training in a later section in this chapter.)

Red Raspberries: While commercial growers often plant red raspberries in hills, rows are the most convenient and fruitful method for home growers. Space the plants 2 feet apart in a row and allow them to fill in the spaces gradually with their root suckers. As the plants develop new canes, you'll find that the rows will get wider. You can allow the brambles to spread out anywhere from 1 to 3 feet. A 3-foot row gives more berries for the area, but you'll have to reach way inside to pick some of the berries. This is fine with a thornless variety, but it might be painful with a thorny one!

Black and Purple Raspberries: These plants don't produce root suckers like red raspberries, so they don't fill in the rows. Plant them in hills 3 to 4 feet apart, to allow plenty of room for their branching canes.

Blackberries: How you plant blackberries depends on whether you're growing an erect, semierect or trailing variety. Plant either of the first two types like red raspberries, for they will sucker from the roots and fill in the row. Trailing blackberries often do not sucker, but their canes are very vigorous and can grow quite long. You can plant these 6 feet apart near the northern edge of their range since the plants won't grow as vigorously. In areas where these plants thrive, give each of them 8 to 12 feet of growing room.

SELECTING BRAMBLES

When you're shopping for brambles, it's very important to select certified disease-free stock. Even if disease-free plants cost more, the investment is worth it. Remember that you're planning to keep your berries going for six to ten years or even longer. If you start out with clean, healthy plants and take good care of them, you will be well rewarded by abundant crops for many years. You're running a risk by taking plants from a friend or by ordering from a nursery that doesn't guarantee disease-free stock. You may end up not only losing your planting but also contaminating the soil and ruining it for years to come in terms of growing other susceptible crops.

We realize, however, that there may be times when you can't resist a bargain, or you become smitten with the flavor of your friend's berries and don't know where to obtain them commercially. Before you move these free plants into your garden, take the time to observe the plants for signs of disease. To be on the safe side, watch them grow for a year before you make the move. Signs of diseased plants to watch for are crumbly berries (in raspberries); leaves that are spotted, yellowed or wilted; white mold on leaves or canes; and purple streaks on canes or leaves. If you don't spot any of these symptoms, and the plants exhibit dark green leaves and vigorous growth, chances are good they're healthy.

Both of us have gotten raspberry plants from friends who take good care of their gardens, with mostly good results. However, not every tale of plant adoption has had a happy ending. Dorothy had to destroy her Fall Gold berries because they were infested with viruses, and Diane managed once to import cane borers with some free raspberries. The borers then proceeded to infest all the plants she put in that bed, not only the free ones. Fortunately, she was able to eliminate the problem by relocating her raspberries. But let that be fair warning to those of you who are tempted by an offer of free plants!

GETTING DOWN TO PLANTING

If your soil is in poor shape, you should try to plan ahead and get it ready for planting about a season ahead of time. When the soil's too heavy, it tends to stay soggy after watering, making your brambles candidates for verticillium wilt and various root diseases. The other extreme, sandy soil that drains too quickly, leaves bramble roots high and dry and susceptible to water stress (75 percent of their roots lie in the top 12 inches of soil). The best way to solve either problem is by making massive infusions of organic matter into the planting hole. As long as you're planning ahead, a green manure crop is a good way to increase both the fertility and the texture of the soil. Sow clover, cereal rye or soybeans the fall before putting in your brambles. Just before the crop flowers, dig or till it into the soil. If the soil is clayey, you might want to dig some coarse builder's sand in at the same time to help improve drainage.

Another way to build up your soil in preparation for planting is to add well-rotted cow or poultry manure. Dig in cow manure at the rate of ⅓ to ½ pound per square foot and poultry manure at ¼ to ⅓ pound per square foot.

If you order brambles by mail, they'll arrive as unpromising-looking sticks a few inches long, with thick, fibrous root systems packed in moist peat moss or

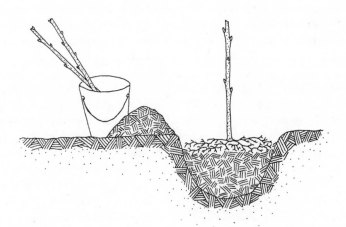

Planting Time: Keep brambles moist in a bucket of water until you're ready to move them into the ground. Prepare a hole that's one to one and a half times the size of the root mass and mix in lots of organic matter. Fan out the roots as you position the plant. Set the bramble at the same depth it was planted in the nursery.

wood chips. (Some nurseries send them rooted in a pot, rather than bare rooted.) It's a good idea to soak bare-rooted plants for a couple of hours in a bucket of water before planting. Take the bucket out into the garden with you when you plant so the roots can be kept wet until the moment they're put into the ground.

For each plant, dig a generous hole one to one and a half times larger than the mass of roots. Fan out the roots as you position the plants. Fill the hole with loose soil to the level at which the brambles were planted in the nursery—indicated by the brown soilline around the canes. If your soil is loose, plant a bit more deeply so that water won't wash the soil away from the roots. After planting, water well, and then water conscientiously every two or three days for the first week.

If your soil is soggy and your climate wet, you can overcome these potential pitfalls somewhat by planting your berries in raised hills or or in a raised 3-foot-wide bed. Mound the soil 6 to 8 inches high and plant one vine per hill or plant as you would in an unraised bed. These elevated areas encourage good drainage. In fact, the drainage is so good that the hills or raised bed may dry out in hot weather unless you mulch them well with an absorbent material like well-rotted sawdust.

TENDER LOVING CARE FOR BRAMBLES

After your plants have been in the ground for a few years the rows may begin to look crowded. This is the time to thin out your brambles. Thinning is best done in the springtime. The desired spacing depends on the type of plant you have. When you thin, always leave the thickest canes and remove the thinnest. (Remember that the thicker the cane, the more berries it will produce.) With suckering types, remove suckers rather than canes from the original plants when you must thin, for the roots of the suckers will compete with those of the main plants for nutrients.

Red and Yellow Raspberries: Each two-year old cane has two underground buds that give rise to new canes in the springtime. A simple mathematical progression shows that the second year you should have two canes from each original plant, the third year you should have four, and the fourth year, eight. You may actually end up with fewer canes, since not every bud will necessarily produce a cane. You will also have root suckers that grow up between the plants and to the sides, widening the row. Because some of the canes grow in clumps from the original plants, and others singly from root suckers, we can't give you an exact spacing for the canes. As a rule of thumb, there should be no more than eight canes for every 3 feet of a narrow row. Generally speaking, the canes should average no closer than 4 to 6 inches apart.

Black and Purple Raspberries: The number of canes you leave in each hill depends on your soil. If it's rich and the canes are growing vigorously, you can leave ten canes per hill. If the soil is very poor, allow only five.

Erect Blackberries: Since these plants have root suckers like red raspberries, they will gradually fill up the spaces between plants. You can allow five to six canes on the average for every foot of wide row.

Western Trailing Blackberries and Dewberries: While the roots of these berries don't sucker as extensively as the erect varieties, now and then suckers will appear. Once you spot any, remove them. Loganberries are very vigorous and will produce many canes each year. Limit each hill to 10 to 14 canes, depending on their vigor. Youngberries will only produce 2 or 3 new canes, so they don't need to be thinned out.

WATERING, FEEDING AND OTHER CARE

Brambles are especially vulnerable to water stress during blossoming, fruit ripening and flower bud development in the fall. All brambles need plenty of water because of their shallow roots, but this doesn't mean you should go overboard. These plants should never be allowed to sit in soggy soil. Brambles benefit especially from drip irrigation, for the ripening fruits tend to soak up water from droplets that fall on them from rain or an overhead sprinkler. This dilutes the flavor of the berries.

Don't assume that when you added manure to the planting hole, your brambles were set for life. Continue to add the same quantity of well-rotted cow or poultry manure that we recommend under Getting Down to Planting to the bramble bed each year. Because the plants' roots are so shallow, don't dig the manure into the soil. Instead, move aside the mulch and spread the manure on the soil surface. The nutrients will make their way down into the root zone and your plants' roots will remain unmolested. Always be sure to add the manure before the plants start growing in the spring.

One note of caution on using manure is in order. While all fruits are sensitive to salts in the soil, raspberries and blackberries are especially so. Salty soil can kill red raspberries and thornless blackberries outright. Excessive amounts of manure (poultry manure, in particular) can lead to salty soil. If you stick to our guidelines for manure applications and stay away from excessive amounts of poultry manure, you shouldn't have any problems, unless your soil was salty to begin with. If soil in your area tends to have this problem, you might want to consider growing these salt-sensitive plants in hills, as we describe under Problems with Fruiting Plants in chapter 1.

Weeding is especially important with brambles, for encroaching plants compete with the shallow bramble roots for nutrients and water. Diane found out just how stiff this competition can be when she first planted raspberries. She dug up a foot-wide row in a grassy part of her yard and never bothered to clear out the area any more; she just kept the grass neatly mowed. When three years went by with no raspberries to eat, she finally realized that her plants were too busy struggling against the stronger grass to bear fruit. After relocating the planting to an open, weed-free area, she began getting generous harvests from those same plants in a couple of years.

Mulching is another key to success with brambles. One especially successful grower we know has kept a thriving raspberry patch going for more than ten years by regular thinning, careful weeding and use of a rotted sawdust mulch several inches thick all around her plants and spread over the path between the rows. Because her plants are so healthy and prolific, she often needs help keeping the berries harvested. Picking berries at her place is a pleasure, for the soft mulch gives gently under our feet, and to pick from the thornless canes is a real joy.

TRAINING BRAMBLES

The easiest way to keep raspberries and erect and semi-erect blackberries in line is by installing a sturdy 4- to 5-foot-tall post at each end of the row. Place a crossbar at the top of each post and another at 3 feet, and string heavy wires between the

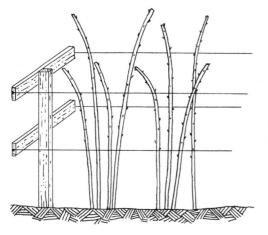

Trellising Your Brambles: Your raspberries and blackberries will be healthier and more productive if you keep them from drooping along the ground. At each end of the row, erect a post with crossbars. Run wire between opposite crossbars to create a sturdy support for the berry-laden canes.

crossbars along either side of the row. The width of the crossbar depends on the width of the row, and the height of the wire depends on the type of berry you are growing. A 4-foot-high top wire is best for black and purple raspberries, while a 5-foot-high one is better for red and yellow raspberries.

As the canes become laden with ripening berries, they tend to fall over and lean against the wires. With a few varieties, such as the short, stocky red raspberries Boyne and Newburgh, and fall-bearing red and yellow types, you might be able to get away without support wires. But most varieties, such as Latham and Canby Thornless, will fall over right to the the ground as soon as the first berries are ripe. Should you be tempted to leave the canes there until it's time for picking, keep in mind that slugs and other pests are likely to get the first bite of such grounded berries.

Trailing blackberries grow so vigorously that they need to be tied to a trellis. Position posts at 15- to 20-foot intervals along the row and string a wire at 3 feet and another at 5 feet. There are various ways of training the vines along the trellis,

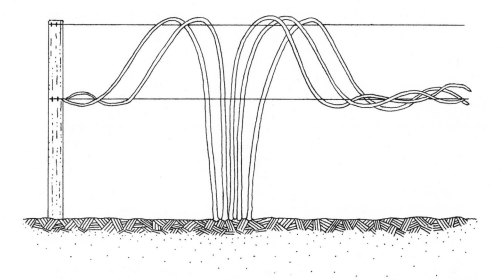

Trellis Tactics for Trailing Blackberries: The lengthy canes on these plants need to be tied to a trellis to keep them under control. On an extra-vigorous plant, loop the canes over the top wire and fan them out along the lower one so they receive as much sun as possible.

but it's best to fan out the canes when you tie them. That way, each cane will get plenty of sunshine and produce lots of berries. If the vines are very vigorous, you can loop them over the top wire and then fan them out along the bottom one.

In mild-climate areas, you can tie blackberry primocanes to the trellis as soon as they're long enough. This will improve the yield the following year by exposing the canes to more sun than they would get lying on the ground. However, if you live in a climate where the vines must be protected over the winter, it's best to let the primocanes grow along the ground. You can then cover them with leaves or straw for the winter and tie them to the trellis in the spring, before they begin to leaf out.

FEARLESS PRUNING

Because their growth patterns differ, the various brambles require different pruning practices. They all have one thing in common, though; the old canes should be cut out down to the ground after they have finished bearing. This will open up the plants to sunlight, allowing the new primocanes to grow more vigorously. Since pests and diseases tend to hide out in the old wood, you must burn the old canes or otherwise dispose of them away from the garden.

Single Crop Red and Yellow Raspberries: Many gardening books advise pruning single-crop varieties back to 4 feet in late winter, before the canes bud out. By following this practice, however, you may be drastically reducing your next harvest of berries, for the most productive sections of the cane tend to be those above the 3-foot level. When you cut back a 6-foot cane to 4 feet, you aren't just removing a large percentage of the fruiting wood; you're also removing the part that would produce the most berries! Commercial growers in Oregon often don't top the canes at all. Instead, they weave them through the wires of a trellis so that even though the canes are long, the berries are within arm's reach. Canes taller than 6 feet are difficult to pick from when not worked into a trellis, so home growers might as well top them at this height.

There's one consideration that may entice you to prune your red and yellow raspberries back pretty far—and that's a desire for really big berries. Severely pruned canes produce larger berries than long canes. Keep in mind, though, that the increased size of individual berries doesn't make up for the reduced yield you'll get.

Fall-Bearing Red and Yellow Raspberries: As we explained earlier under How Brambles Grow, fall-bearing raspberries give you two crops a season. The first is produced on the floricanes, like standard varieties, and the second comes later on, on the top part of the primocanes. With these varieties, the tops of the primocanes should be pruned after bearing is over or after frost has ended the harvest. In areas where a long fall ensures an abundant crop, you can cut down all the canes in the fall. The next season, you end up foregoing a summer crop on the second-year canes, but your reward is a heavier fall crop, since the young canes don't have to compete for light and nutrients with the older ones.

Black and Purple Raspberries: While the canes of red and yellow raspberries don't branch the first year, those of blacks and purples do, especially if you encourage them. And you should encourage them, since the more branches the plant has, the more flowers (and therefore berries) there will be. The apical tip at the end of the cane produces auxin, which inhibits the vine branches from growing. By pruning off the tops of the primocanes, you remove the source of auxin, and branches can grow freely. Don't be surprised to see branches grow as long as 5 feet! The next year, these branched canes will produce much more heavily than unpruned, unbranched canes.

Here are some measurements to guide your pruning. When the primocanes reach 18 to 24 inches in height, cut off the top 3 to 4 inches. Late the next winter, you should cut the lateral branches back to 8 to 12 inches, which will encourage them to produce more fruiting branches.

Upright Blackberries: You can prune these like black raspberries, cutting the primocanes back to 30 to 36 inches and the lateral branches back to 12 inches. In a few blackberry varieties, the flowers are borne far out on the ends of the laterals. If you have such a variety, don't cut the laterals back very far.

Western Trailing Blackberries and Dewberries: During the first year, set your pruners aside and let these berries grow freely. Then, in late winter, prune the canes back to about 10 feet in length. In areas where the growing season is nice

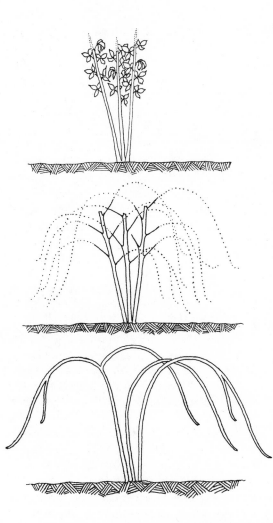

The Right and Wrong Way to Prune: With black and purple raspberries, the more branches there are on a plant, the more berries you'll harvest. In the summer, prune the top 3 to 4 inches off each primocane, as shown at top. Late the following winter, prune lateral branches back to 8 to 12 inches. At bottom is a plant that was ignored and never pruned. The unbranched canes will produce some berries, but not nearly in the same abundance as pruned, branched canes will.

and long, all the canes of dewberries are sometimes cut back to the ground after harvest. The plants are still able to bear a crop the next season, and pests and diseases seem to bother them less when they're treated in this way.

PROPAGATING BRAMBLES

Brambles propagate readily, so it's really quite easy to get new ones going if you want to increase the size of your planting or move it to another location. With red and yellow raspberries and upright blackberries, you can carefully dig up root suckers and move them to a new location. Take suckers no closer than 6 inches to the mother plant and dig deeply so you're sure to get them out intact. Try to do this while the plant is still dormant. If you can't, you can still move suckers within the first month after the plants have started growing—but don't delay any longer. The longer you wait, the greater the chance you'll damage the mother plant.

Black and purple raspberries, thornless blackberries and trailing varieties must be propagated by tip layering. You should do this late in the season—early August in New York and early September in the southern states, for example. Bend down the first-year canes and bury their tips 4 inches deep in loose soil. Roots will eventually grow from these tips. In the spring, cut off the tip of each cane leaving an 8-inch stub on the end that's in the ground, and dig up the new plants. Leave a generous ball of soil around the roots and transplant them to their new locations.

Tip Layering: It's easy to enlarge your bramble patch when you're growing black and purple raspberries, thornless blackberries and trailing varieties. Just bend the tip of a first-year cane down to the soil and bury it about 4 inches deep. This tip will sprout roots in several months. Cut the new plant away from the parent plant and carefully transplant it to its new home.

HARVESTING AND STORING BRAMBLE BERRIES

You might find it interesting to know that the weather can affect the flavor of at least one of the bramble fruits. For the best-tasting raspberries, you should always hope for sunny weather during ripening. Cloudy, wet weather inhibits the plant's ability to take up nitrogen from the soil. This is significant because nitrogen is an important component of the organic acids that help give raspberries their special, intense flavor.

How can you tell when berries are at their peak of sweetness and juiciness? Raspberries should have an even color all over their surface. They're ready to pick when they come off the plant at the slightest touch. Blackberries, too, should part from the plant easily when ripe. They are at their sweetest just after they lose their shininess and become a bit dull in color.

Freezing is a good way to capture summer's sweet berry harvest and hold it over for later enjoyment. Bramble fruits freeze very easily; just spread them out on cookie sheets, slip them into the freezer, and once the berries have frozen hard, store them in heavy plastic bags. They don't keep well in the refrigerator, however, so don't delay your freezing or jam making longer than a day or two after picking. Once they're ripe, bramble berries don't stay in good shape for very long in the garden, so check the plants at least every other day to avoid rotten and fallen berries.

PROBLEMS WITH BRAMBLES

Brambles are relatively trouble-free plants. Their canes should remain healthy as long as you practice some preventive measures. Cut back old canes as soon as they bear; keep down weeds; don't allow the rows to become too dense; and quickly remove any sick-looking canes. Various species of borers can infest the canes, but if you cut out and dispose of any wilted-looking canes right away, you should be able to avoid trouble with these pests.

Viruses are probably the worst problem you'll encounter with brambles. Black raspberries and certain varieties of red and yellow raspberries (Fall Gold and Latham, in particular) are more susceptible to viruses than others. Clues that your plants may be infected are yellowing foliage, poor growth, and crumbly berries.

Crumbly berries by themselves, however, can be caused by a number of conditions. One of these is poor pollination. The individual druplets that make up the raspberry have tiny, microscopic hairs that help hold them together. Every red raspberry flower has 100 to 125 pistils that can be pollinated and develop into druplets; however, usually only 75 to 85 actually do. If many fewer than this form, the fruit will crumble, since there aren't enough druplets to cling well to one another. Rainy or cloudy weather during blooming can cause more crumbly berries than usual to form, since these conditions hamper pollination.

Other factors responsible for crumbly berries are water stress; a soil deficient in phosphorus, nitrogen or iron; or exposure to volatile herbicides such as the common lawn weed killer 2, 4-D. If none of these problems exist in your garden but your raspberries do crumble, disease is almost certainly the culprit. Then you must dig up the entire row, dispose of the plants (never in the compost pile), and plant fresh virus-free stock as far away from the old planting site as possible.

Verticillium wilt, a fungus disease, is another potential spoiler in the bramble bed. On black raspberries, the lower leaves take on a yellow-bronze color, then curl upward. Eventually they turn brown and fall off. The canes may have blue or purple streaks, and may even die before the fruit matures. Red raspberries are generally more resistant, but they too can succumb. Symptoms in these plants are yellow-bronze leaves that curl downward. See Problems with Fruiting Plants in chapter 1 for a more detailed discussion of this disease.

Wasps or hornets sometimes develop a great attraction to raspberries. They alight on the berries and suck them dry, druplet by druplet. Not only can you lose a portion of your crop to these alarming pests, but they can make you quite reluctant to venture out and pick your berries. If you find that wasps or hornets are cheating you out of a harvest, consider setting up fly traps baited with bits of tuna, liver or sweet fruit punch. These will lure the pests away from your berries. Be sure to pick all the ripe berries, even those past their prime that you won't use, so there's nothing there for the wasps to eat. The cool early morning is the best time for harvesting, since the wasps won't be out yet. But be careful! They may have stayed the night on the berries.

BRAMBLE FRONTIERS

As with other fruits, breeders are always working on developing varieties as resistant to disease as possible. One way raspberry breeders identify disease resistance is to grow their experimental varieties in fields infected with disease organisms. That way, any vigorous canes that develop are sure to be disease resistant!

New kinds of two-crop raspberries are also in the developmental stages. Because they can be grown as annual crops, such berries are attractive to commercial growers. But home gardeners have reason to welcome more two-crop varieties, as well. These plants are time-savers since they don't need to be trellised, and they effectively extend the raspberry season since they produce two crops each year.

Guide to Bramble Varieties

BLACKBERRY VARIETIES

Black Satin—Large, midseason. Zones 6–9. Trailing, thornless, disease resistant, widely adapted.

Boysen (Boysenberry)—Large, purple, excellent flavor. Zones 7–9. Upright.

Darrow—Large, glossy black, midseason to late. Zones 5–8. Upright, vigorous, excellent producer.

Ebony King—Glossy purple, tangy flavor, early. Zones (4) 5–8. Upright, very hardy.

Evergreen—Large, sweet, midseason. Zones 6–9. Thornless, drought resistant.

Logan (Loganberry)—Large, slightly tart, flavorful, late. Zones (6) 7–9. Trailing, does well on West Coast, not hardy in East.

Lucretia—Extra large, long, early. Zones 7–10. Trailing, disease resistant, does well in Southeast.

Marion—Medium to large, early to midseason. Zones 7–9. Trailing, does well in Oregon and Washington.

Rosborough—Extra-large, juicy, early. Zones 7–9. Erect, very vigorous, does well in South.

Smoothstem—Medium-large, late. Zones 7–8. Semi-erect, thornless, productive.

Thornless Boysen—Extra-large, midseason. Zones 5–9. Trailing.

Young—Large, sweet, few seeds, midseason. Zones 7–9. Trailing, does well in West and along Gulf Coast.

RED, ONE-CROP RASPBERRY VARIETIES

Boyne—Medium, midseason. Zones 3–8. Vigorous and hardy.

Canby Thornless—Medium to large, firm, outstanding flavor, early. Zones 4–8. Good for freezing.

Dorman Red (Dormanred, Dorma Red)—Extra-large, early. Zones 7–10. Extremely vigorous plants, best variety for Deep South.

Latham—Extra-large, midseason. Zones (3) 4–8. Productive, adaptable to many locations. Prone to virus.

TWO-CROP RASPBERRY VARIETIES

Fall Gold—Large, yellow, early and late. Zones 4–8. Prone to virus.

Fall Red—Large, bright red, early and late. Zones 3–8.

Heritage—Extra-large, red, mild flavored, early and late. Zones 4–8. Upright canes.

Southland—Red, good tasting, early and late. Zones 5–8 (9). Disease resistant.

BLACK RASPBERRY VARIETIES

Blackhawk (Black Hawk)—Large, firm, very good quality, early. Zones 5–8. Vigorous, tolerant of anthracnose.

Bristol—Large, excellent quality, midseason. Zones 5–8. Vigorous, very productive.

Cumberland—Large, midseason. Zones 5–8. Reliable, bears well, susceptible to fungus and virus.

PURPLE RASPBERRY VARIETIES

Brandywine—Large, firm, tart, late. Zones 4–8.

Royalty—Large, late. Zones 4–8. Resists aphids; less susceptible to virus. Can pick early while still red; immature berries taste like red raspberries.

CHAPTER 4

BLUEBERRIES
A BOUNTY OF BERRIES

Vital Statistics

FAMILY
Ericaceae

SPECIES
Vaccinium angustifolium (lowbush)
V. ashei (rabbiteye)
V. corymbosum (highbush)

POLLINATION NEEDS
Highbush will set more fruit with
cross-pollination. Lowbush and
rabbiteye require cross-
pollination.

HOW LONG TO BEARING
3–4 years

CLIMATE RANGE
Lowbush   Zones 4–7
Rabbiteye   Zones 7–10
Highbush   Zones 4–8 (marginal in
Zone 3)

The word blueberry brings to mind mouthwatering images of creamy cheese-cake topped with plump, juicy fruit, or of luscious, lustrous jam. Despite the popularity of this special fruit in the kitchen, blueberry varieties for the home garden have been developed only relatively recently and are not as widely planted as they deserve to be. We hope to change that by telling you just how carefree and prolific these berry bushes can be.

Blueberries have a lot going for them. Once established, they take little care and attention. Without being fussed over they can yield a generous crop of delectable

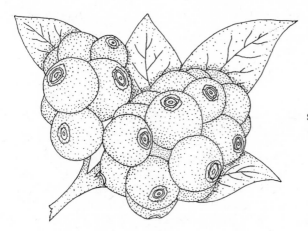

The Beauty of Blueberries:
Fresh, plump blueberries
straight from the bush are a taste
treat gardeners in almost every
part of the country can enjoy.

berries far sweeter and far less expensive than any you can buy in the store. In addition to their delicious flavor when fresh, blueberries freeze especially well. This allows you to enjoy a touch of summer in the dead of winter by preparing melt-in-your-mouth blueberry muffins or steaming stacks of pancakes bursting with plump, juicy berries. Blueberry bushes, with their flaming red foliage, are also very decorative in the fall. Diane grows them as a border along her driveway and around her lawn, where they provide a beautiful autumn frame for the meadow and forest beyond.

SOME BLUEBERRY BACKGROUND

Blueberries are North American natives, so they're attuned to our seasons. They're in the same plant family as azaleas, rhododendrons and heather, and share the same genus as cranberries. There are three basic types of blueberries—lowbush, highbush and rabbiteye. Here's a brief character sketch for each of these types.

LOWBUSH BLUEBERRIES

Lowbush varieties (*Vaccinium angustifolium*) were probably the first to be managed by humans. Indians in the New England area found that burning the native lowbush blueberry plants would keep them productive by destroying the weeds that grew among them. The bushes survived the burning because of their special underground stems. Early settlers followed the Indians' lead and continued this practice.

Roughly 200 years later, the commercial industry harvests more lowbush blueberries than any other type. But lowbush blueberries aren't grown much outside of their native area because, until recently, it has been difficult to propagate them. Those difficulties have been overcome, and now lowbush varieties for home gardeners are being developed. Keep watching for signs of these varieties in your nursery catalogs as they become more widely available.

HIGHBUSH BLUEBERRIES

Highbush blueberries (*Vaccinium corymbosum*) are the kind most commonly grown in home gardens. Selective breeding of highbush varieties began in 1909, when Dr. F. V. Coville started looking for varieties with large berries for commercial use. He made crosses involving two species of highbush blueberry and some lowbush species. (Some of the varieties with lowbush parentage are June, Weymouth and Earliblue.) Because many blueberry species are cross-fertile, Dr. Coville had a large range of genetic characteristics to choose from in developing new varieties. By the time of his death in 1937, he had produced 15 improved blueberry varieties. Other researchers have been busy developing blueberries since then, and presently there are more than 45 varieties available, with new ones being introduced often.

Highbush blueberries grow from 4 to 10 feet tall, depending on variety and locale. The growth habits of the various varieties can be quite different, so you should check the descriptions carefully if you're planning on planting a hedgerow. It takes a few years for the plants to reach maximum production, but it's worth the wait. A mature six- to ten-year-old bush will produce 4 to 8 pints of delicious berries in a season, while a plant that is ideally located and given exceptional care can yield over 25 pints.

RABBITEYE BLUEBERRIES

Gardeners south of the Mason-Dixon line shouldn't despair, for they can still grow blueberries. Rabbiteye varieties require only one-third as many cold hours as the highbush types, suiting them perfectly to the Deep South.

The rabbiteye blueberry (*Vaccinium ashei*) is native to Georgia, southern Alabama and northern Florida. It's easy to understand, then, why this plant isn't as hardy as the other types, and why it can't be grown farther north than Maryland. Because it has a different number of chromosomes, the rabbiteye blueberry is very difficult to cross with either highbush or lowbush types. This is unfortunate, for rabbiteye bushes can grow in drier soils than highbush or lowbush plants and are more productive because of their larger size—desirable traits that breeders would like to introduce into other species. Berries from wild rabbiteye plants aren't very promising breeding material—they're small, black, gritty and practically flavorless. But rabbiteye breeding programs are enthusiastically underway, with more than ten varieties that bear large, tasty berries now available.

BLUEBERRY NUTRITION

Along with their sweet juicy flavor, blueberries provide some impressive nutritional benefits. Highbush blueberries have a high vitamin C content—anywhere from 14 to 27 milligrams per 100 grams of fruit, about a half cup of berries. (Lowbush varieties aren't quite as rich in vitamin C.) On top of that, blueberries are a good source of iron; 1 cup of raw berries contains 1.5 milligrams. All this nutritional goodness comes at a real calorie bargain—approximately a half cup of berries has only 55 calories.

HIGHBUSH VARIETIES

Because they're so closely related, the three types of blueberries have many traits in common. However, they also differ in significant ways. We'll cover highbush berries, most commonly grown in home gardens, here in detail, and discuss the special properties of lowbush and rabbiteye varieties later in the chapter.

Highbush blueberries vary greatly in their disease resistance and hardiness, with particular varieties developed for different regions of the country. For example, southern varieties have been bred for resistance to the disease cane canker. Northern varieties lack this resistance, so if you try to grow them in the South, they're likely to become infected.

As with other fruits, blueberries require a certain amount of chilling during the winter to break dormancy, leaf out in the spring, and blossom. In general, they must undergo 650 to 850 hours below 45°F. Still more chilling hours will improve

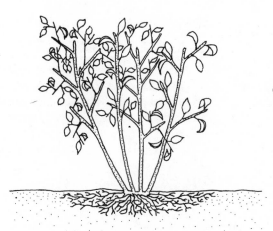

Highbush Blueberry: The most common type found in home gardens, this blueberry grows from 4 to 10 feet tall. The oldest canes will have the smallest leaves while the newest ones will have the largest leaves. Compare the leaves of the four-year-old cane on the left with the leaves of the one-year-old cane on the right.

flowering and foster early bud break. This chilling requirement varies from variety to variety and is often the determining factor when it comes to selecting plants. Varieties developed for the North won't undergo enough cold to break dormancy in the South. Southern varieties, in addition to not being as cold hardy, may break dormancy early in the North and then suffer frost damage during cold spells. In general, blueberries need to have a growing season of 120 to 160 days between killing frosts (20°F).

In the North, blueberries will survive best during a consistently cold, dry winter. High humidity lowers their hardiness, and warm spells can make the plants jump the gun and begin to break dormancy. Once this happens, the plants are much less cold hardy.

The relationship between humidity and hardiness can be quite striking. For example, researchers at the University of Minnesota found that an unnamed experimental highbush variety had flower buds hardy all the way down to −40°F when the humidity was low. But buds of the same plant were only hardy to −5°F when the relative humidity was between 96 and 100 percent.

Although you can't do much about the humidity level, you do have control over which varieties you plant. Selecting hardy varieties tips the odds in your favor that the plants will make it through the winter. Listed in decreasing order of hardiness, northern highbush varieties rank as follows: Northland, Bluecrop, Blueray, Herbert, Jersey, Burlington, Rubel, Earliblue, Rancocas, Weymouth, Atlantic, Collins, Berkeley, Coville, Pemberton, Dixie, Stanley and Concord.

For southern highbush blueberry growers, hardiness isn't as important a consideration as it is for their northern counterparts. Since the southern winter is milder than the northern one, they need to find varieties with low chill requirements. Some good southern varieties that break dormancy after relatively few hours' exposure to temperatures below 45°F are Angola, Croatan, Murphy, Scammell and Wolcott.

Matching the right variety to your particular region does more than ensure the survival of the plants. It also affects the quality of your berries. The size of the berries themselves appears to be influenced by winter's cold. After a warm winter, blueberries are smaller than after a cold winter, so it appears that adequate chilling is necessary for the fruit to develop properly. This is yet another reason not to grow northern highbush varieties in southerly climates. To make sure you choose the best variety for your area, refer to the Guide to Blueberry Varieties before you buy your plants.

THE UNIQUE ROOTS OF BLUEBERRIES

Blueberries differ from other cultivated fruits in one very important respect—their roots. In their natural habitat, blueberries live in very moist areas. Often, the water table is less than 2½ feet deep, and the soil is rich in organic matter. In this kind of environment, the roots have an easy job getting enough water and nutrients. Because of this easy access, the root system doesn't need to be an extensive, far-reaching one. If you compare blueberry roots with those of any other fruit, you'll see how seemingly underdeveloped they are. While most plants have branching roots that probe deeply through the soil in search of water and nutrients, blueberries have a shallow mat of fine, fibrous roots. These lie mostly within the top 14 inches of soil and rarely extend beyond the drip-line. It's a good thing they don't have to form an extensive system since blueberry roots grow very slowly.

Another curious fact about blueberry roots is that they lack root hairs. This drastically reduces their total surface area. Just how drastically becomes apparent when you compare a foot-long section of a blueberry root with a piece of wheat root the same length. The wheat root has ten times as much surface area as the blueberry root!

It's important for you to remember these quirks that blueberry roots have. Because they're so shallow and limited in the area of soil they can mine, you need to be especially attentive to their water and nutrient needs.

FRIENDLY FUNGI

Like many other plants, blueberries form associations with mycorrhizal fungi. This is a friendly arrangement, for the fungi help the plants take up nitrogen and phosphorus from the soil. This is especially important for blueberry plants, since their roots lack root hairs.

Studies have shown just what a difference these fungi can make in the overall health and well-being of blueberry plants. When growers in New Zealand first planted blueberries, they found to their dismay that the plants grew poorly. Scientists suspected the reason these North American natives weren't doing well was that New Zealand soils lacked the proper mycorrhizal fungi. To test this hypothesis, they grew six different blueberry varieties. Some of the bushes were inoculated with the appropriate fungi and others were not. After four years, the scientists found that the inoculated plants of every variety produced significantly better than the non-inoculated ones—sometimes strikingly better. The variety Stanley produced 92 percent better when inoculated; Blueray, 69 percent; Ivanhoe, 18 percent; Herbert, 51 percent; Jersey, 39 percent; and Dixie, 11 percent.

THE REASONS BLUEBERRIES NEED ACID SOIL

You've probably read that blueberries require especially acid soil and may have wondered why. Actually, acidic soil is only one part of a chain of events that is aimed at blueberry roots getting enough iron. Iron is an important nutrient for vigorous plant growth. Before iron can be taken up by roots, however, it needs to be brought into solution in the soil. In order for this to occur, an acid condition is necessary. Plants that don't need acid soil have roots that can secrete their own acid. The acid then reacts with the iron, making it available to the plant.

Blueberry, azalea and rhododendron roots can't do this, so the soil has to be acidic. If there's not enough acid present, the leaves on these plants will turn yellow from lack of iron, and the plants will languish. Interestingly enough, some blueberry hybrids combining highbush, lowbush and rabbiteye species are able to release acid and can grow in more neutral soils. With any luck, such varieties may someday be available for home gardeners.

Iron isn't the only nutrient affected by soil acidity. A low pH also enables blueberries to extract nitrogen more easily from the soil. Nitrogen is available in several forms within the soil; blueberries have an easier time assimilating nitrogen as ammonium than they do nitrogen as nitrates. Deep within the ground, there are two types of soil organisms working at cross-purposes. One converts nitrates into ammonium and the other converts ammonium into nitrates. Whichever one is dominant is determined by the soil pH. At a higher pH, the soil organisms that convert ammonium into nitrates are more active, and more of the soil nitrogen is present in this less accessible form. But if you take steps to lower the pH, the other organisms gain the upper hand, and more ammonium is present.

SELECTING A SITE

Since blueberries have such fine, slow growing roots, they do best in loose soil rich in organic matter. Not only does organic matter allow the fine roots to penetrate easily, but it also holds moisture like a sponge. Readily accessible soil moisture is especially important to blueberries because their shallow, hairless roots can't plumb the depths for water. You should keep all these considerations in mind when you choose the site for your blueberries.

If your soil is loaded with organic matter and has a pH of 4.5 to 5 (the best range for highbush plants—for the other two, see the sections on Rabbiteye and Lowbush Blueberries later in the chapter), you will have no trouble growing blueberries. Unfortunately, not everyone will be so lucky. Sometimes the soil in a yard will vary radically from one spot to the next. If your soil varies, use a home soil test kit to determine the most acid area for planting in your yard.

There is one last consideration in choosing your site—avoid planting blueberries in an area where they're likely to be disturbed. The canes are very brittle and break off easily, so they'll really take a beating if they're subjected to a lot of wear and tear. Both of us have learned this lesson the hard way. Dorothy lost three bushes over the winter because they were planted next to the busy driveway. Active children, snow shoveling and a delivery of wood reduced the plants to nothing. Diane lost many canes one year to a violent dog fight that took place right in the middle of her blueberry patch. There's nothing worse than seeing bits and pieces of your bushes bite the dust, so try to plant your blueberries in a quiet, out-of-the-way part of your yard.

PLANTING BLUEBERRIES

As long as a blueberry plant is dormant, you can plant it in the spring or fall, whichever is more convenient for you. No matter when you plant, you may find that you have to hold your nursery stock over for a few days before you can actually get out to work in the garden. Your first inclination may be to stick your plants in a bucket of water—but don't! We've stressed over and over that blueberries need plenty of water, but they can get too much of a good thing. If you leave plants in a bucket of water for a couple of days, the plants will drown! Instead of placing them in water, sink them into a temporary trench out in the yard. Follow the directions for heeling-in that we give in chapter 1 under What to Do When Your Plants Arrive. If you'll only be holding the plants over for 24 hours or less, you can set them in a bucket of water—but don't tempt fate by leaving them there any longer.

When it comes time to mark off planting holes, be sure to space highbush blueberries at least 6 feet apart—7 to 8 feet between the plants is even better. Space the rows 10 feet apart to give you plenty of room to walk and harvest between rows. As you dig the holes make them a generous size. They should be large enough to allow the roots to spread out in a horizontal mat—they won't do well if they're bunched up. Dig each hole deep enough so that the root mass will lie ½ to 1 inch below the surface of the soil.

If your soil isn't acid enough don't be discouraged. You can provide the humusy, acid soil blueberries need with surprisingly little effort. Mix the soil you've excavated from the planting hole with an equal part sphagnum peat moss. Be sure to use sphagnum peat moss since some other types are less acidic. Use this 50-50 mixture to fill in the hole around the roots of your bush.

This very simple matter of mixing peat moss into the planting hole is really quite effective at modifying soil that isn't as acid as it should be. It doesn't really take that much time and effort when you consider the consequences of *not* doing it at all. In Diane's early blueberry growing days, she was ignorant of the fact that you should compensate for a non-acid soil when you plant. She planted a whole row of blueberries without enriching the soil at all. It just so happens that her row fell along a gradient of soil that ranged from dark, loamy and acid to rocky, sandy and not acid enough. Although Diane has dressed the plants with a peat moss mulch, no amount of hindsight will make up for the fact that she didn't acidify the planting holes in the rocky, sandy soil. The bushes on the end where the original soil was rich and acidic still do much better than those at the other end.

Your blueberry planting task isn't over until you add a 3- to 6-inch layer of mulch. Blueberry plants love mulch. Their shallow roots appreciate the moisture

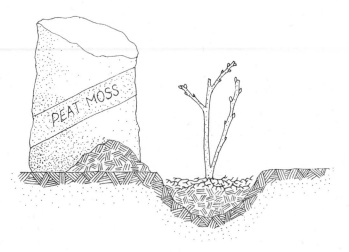

Remedying Low-Acid Soil: Blueberries need acid soil in order to grow well. If your soil isn't acid enough, prepare a 50-50 mix of sphagnum peat moss and soil to fill the planting hole. When you plant, fan out the roots and position the bush so the root mass will be ½ to 1 inch below the soil surface. When you're done, add a 3- to 6-inch layer of acid mulch as a finishing touch.

the mulch holds within the top layer of soil. Without a mulch, these top few inches of soil are more likely to dry out, with the result that the roots may be damaged. A mulch will also discourage weeds from growing in your blueberry patch. Given half a chance, weed roots will rapidly become intermingled with the shallow blueberry roots, competing with them for nutrients and water. When you pull out the weeds, you'll end up disturbing the berry roots. Your best bet is to keep weeds from showing up in the first place.

Mulch also helps keep the soil pH at the right level, as long as you use an acid one. Peat moss makes a good mulch, but you must be sure to keep it moist; once it dries out it can blow away easily, and bone dry peat is difficult to rewet. Pine sawdust is also a good acid mulch, but it must be well rotted. If you use fresh sawdust, microorganisms in the soil responsible for breaking it down will tie up the available nitrogen as they do their work. In short, your mulch will be stealing nitrogen away from your growing plants—a situation that's easy to avoid by letting the sawdust age for a season or two before using it. Cottonseed meal and shredded oak leaves are two other good acid mulches. *All* blueberries will benefit from an acid mulch, not just those planted in more neutral soils.

HOW BLUEBERRIES GROW
—AND HOW TO PRUNE THEM

Highbush and lowbush blueberries share an unusual pattern of growth. The first year, the roots send up straight, unbranched, green canes. Then, the apical tips die and turn black. During the fall and winter, these primary canes change from green to brown. The second year, the primary canes produce lateral branches. (Since the tips have died, there won't be any terminal growth.) Like those on the primary canes, the tips of the lateral branches abort. When the tips stop growing, flower buds start to develop on the branches near the tips. As the third year rolls around, this growth pattern repeats itself. The second-year branches themselves branch, then these new branches stop growing and produce flower buds. On unpruned plants, this pattern of smaller and smaller twigs continues each successive year, and the growth on each branch becomes finer and twiggier. Meanwhile new, strong canes (suckers) develop from the roots each year, beginning the cycle over again.

You must keep this growth pattern in mind in order to understand the basics of blueberry pruning. With many other fruits, pruning involves cutting back the tips of branches. If you were to apply this practice to blueberries, you'd end up with little or no fruit, since the flowers develop near the branch tips! Another thing to keep in mind is that the older a cane is, the tinier the fruiting branches, and the smaller the resulting berries. For these reasons, pruning blueberries means cutting most of the four-year-old canes back to the soilline in the springtime. If you leave even a 2-inch stub behind, new growth may develop there, and it will be weaker than that coming directly from the roots.

A PRUNING TIMETABLE

Blueberry plants sold by nurseries are usually two or three years old. Once you plant, your task is to guide these bushes into the most healthy, productive type of development. Let them grow "as is" the first year, after cutting off any wood that may have been damaged in shipping or any weak, twiggy growth. Generally, young blueberries don't need to be pruned for the first three years after planting. When plants reach the three-year mark, you should get your pruners ready and look over your plants with a critical eye. As a rule, once they're established, blueberries should be pruned yearly, in late winter before new growth begins.

Established blueberry plants grow best if they're allowed to keep one cane for each year of age plus one or two vigorous new canes. If the new growth is weak or the new canes are thinner than ¼ inch in diameter, cut back one of the two new canes to the ground. If any of the branches are diseased or frost damaged, cut these back to healthy wood. Frost damage will show as wrinkled or soft canes with a dark color. The pith of the canes will also be dark brown instead of light in color. When you're selecting which of the four-year-old or older canes to keep and

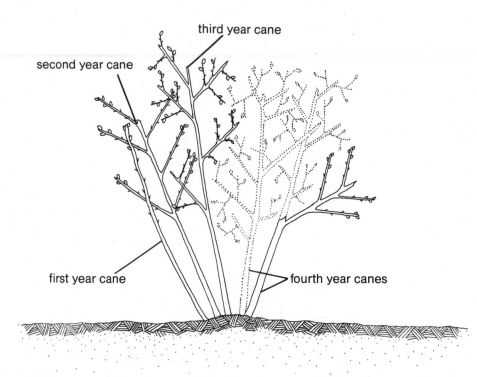

Blueberry Pruning Simplified: What you see is a four-year-old bush after a late winter pruning session. A cane for each year of age has been left, and any diseased and frost-damaged parts have been clipped off. An excessively twiggy four-year-old cane has been cut away, and the twiggiest portion of the remaining fourth-year cane has been cut back to a strong lateral. This illustration has been simplified to show you the basics. On your own bushes you should have one or two vigorous new canes in addition to the ones shown here.

which to prune away, always get rid of the oldest, twiggiest canes. These would produce the fewest flowers, so you really won't miss them when they're gone. If your plants have suffered from extensive winterkill, leave the remaining healthy older canes.

After this basic pruning, cut off the excessively twiggy growth on any remaining four-year-old canes, pruning back to a strong, unbranched lateral. This will ensure that the berries that do develop on that cane will be large. You might find it interesting to know that there's a direct relationship in blueberries between wood thickness and berry size. Branches that are ⅕ inch in diameter produce the largest berries, while wood less than ¹⁄₁₀ inch in diameter yields very small berries.

If your plants start to produce lots of tiny berries that don't taste very sweet, it's a sign they're overbearing. You can get them back on track producing larger, sweeter berries by trimming off some of the weaker, flower-bud-bearing tips. You

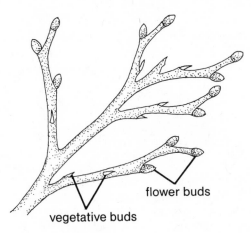

vegetative buds

flower buds

Which Bud Is Which: When you're pruning to encourage larger, sweeter berries, you need to distinguish between flower buds and vegetative buds. Clip off some of the branch tips with round flower buds and leave the thinner, more pointed vegetative buds behind.

can identify the flower buds by their rounded shape; vegetative buds are thinner and more pointed. Flower buds also tend to be located closer to the branch tips than the vegetative ones.

As your plants get older, you should adjust their pruning regimen. Instead of allowing one cane per year of age (which could leave you with a real thicket on venerable bushes), leave only six to eight canes on mature plants, plus two new, robust suckers. This will restrict growth to the most vigorous parts of the bush and will give you larger, more abundant berries.

REJUVENATING OLD, NEGLECTED BUSHES

If your garden has older blueberry bushes that haven't been pruned regularly, you may have to bite the bullet and prune them severely to get them back in shape. Unpruned bushes usually yield tiny berries and produce fewer of them each year. Diane found this out with a Herbert blueberry bush which she neglected to prune. One year she harvested a fine crop from it. The next year the plant also set many berries, but they were much smaller. The third year without pruning, the bush yielded fewer berries, and what berries there were were very tiny and took a long time to pick.

If you find yourself in a similar situation, you'll have to do some remedial pruning. Cut out all the canes older than five years and any others that are weak and thin. Your work will be rewarded in the springtime when strong new suckers emerge from the roots and stronger fruit-bearing branches appear on the younger canes of the plants. That first year the yield may be low, but by the second year your bushes should be well on their way to healthy, generous production.

With *really* old (over 20 years), untended blueberry plants you can try radical pruning. By radical, we mean just that—cut the entire bush back to the soilline.

You won't get any berries the first year, but by the second there will be a small crop. By the third year, the yield will increase dramatically. At this time you should resume pruning as we described earlier for a newly planted bush. In two more years, the yield of these once-neglected bushes will be up to three times the harvest of their previously unpruned state.

KEEPING BLUEBERRIES WELL NOURISHED

As time passes, you should watch your blueberry bushes closely to make sure they're getting the proper nutrition. They'll give you easy-to-read signs if they're not.

If you've had to change the pH of your soil to make it more acid, pay close attention. With time, the pH of a soil that isn't naturally acid can gradually become more basic. This can affect the health of your bushes, as we explained earlier under The Reasons Blueberries Need Acid Soil. Should the leaves on your plants be yellowish when they break forth in springtime, instead of being a healthy dark green, check the soil pH. If the level has indeed crept upward, you should apply a layer of acid mulch to lower the pH. We recommend four good mulches under Planting Blueberries earlier in the chapter. Don't attempt to work these materials into the soil, as you could harm the shallow roots. As the mulch breaks down it will acidify the soil, with no risk to the shallow mat of roots. Be sure to renew this mulch every season.

Soil pH isn't the only thing that should concern you. Blueberries must also receive the right amount of nitrogen. Too little can lead to slow growth, small berries and, in severe cases, reddening of the leaves (this can also be caused by water stress). The other extreme, too much nitrogen, is no better, since an excess will prompt your plants to spend all their energy putting out vegetative growth instead of big blue berries. A good way to give the bushes a well-balanced feeding is to pull back the mulch and lay down a 1-inch layer of well-aged barnyard manure or compost, then pull the mulch back on top. Avoid overdoing it by steering clear of powerful sources of nitrogen such as poultry manure.

Other than soil pH and nitrogen levels, you don't have to worry about your blueberry bushes' nutritional status. Phosphorus and potassium deficiencies are rarely a problem with blueberries.

AN INSIDE LOOK
AT FLOWERING AND POLLINATION

While you're harvesting your berries, very serious business is going on right before your eyes in the blueberry patch—only you can't see it. Even as you're enjoying this season's harvest, the plants are hard at work developing flowers for next year's crop. You can't actually see these flowers—they're locked inside buds, ready to burst forth when their time comes next spring.

Shortening days stimulate the initiation of blueberry flower buds from late August to the middle of September. Miniatures of all the flower parts are formed before the bushes go dormant for the winter. In the springtime, cells forming the egg and pollen grains divide and the flower grows to maturity.

The actual construction of a blueberry flower is interesting to study. Each flower consists of petals that are fused into a tube. The tube can be straight or shaped like a bell, globe or urn, depending on the variety. The flower hangs downward, with the unfused tips of the petals curved gracefully upward. The style hangs out of the flower, while the anthers inside are fused together by tiny hairs.

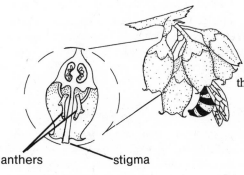

Inside Look at a Blueberry Flower: Blueberry blossoms aren't as showy as those of some other fruiting plants, but they do serve their purpose. A bee must bump into the anthers and get a coating of pollen on its way to the nectar at the base of the flower. When the bee moves on to the next flower, it will brush against the stigma and dust it with pollen.

anthers   stigma

The blueberry flower has a clever set-up that makes sure that pollinating insects will get a liberal coating of pollen. The nectar is produced at the base of the flower, so bees can't get at it without bumping into the fused anthers. When they're jiggled, the anthers all release heavy, sticky pollen that clings to the bee's body. Self-pollination is almost impossible, for the tip of the style is an inverted cone. Any pollen from that flower will hit the sides of the style and roll off without touching the receptive stigma on the tip.

Within the ovary, up to 90 small egg cells wait to be fertilized. However, only 20 to 55 percent of them will be successfully fertilized and go on to develop into seeds. As with other fruits, the more seeds that set, the larger the mature berry will be. Don't be misled by the small size of the fruit—it takes a lot of pollen to fertilize one berry since each pollen grain can fertilize only one egg. What could be a potential problem is overcome very nicely by the arrangement of the various reproductive organs inside the flower. Because the stigma has such a broad surface, a nectar-hunting bee can't get to the nectar without brushing against it and dusting it with the pollen on its body.

Blueberry plants need plenty of help from bees in order to produce a good-size crop. While apple or peach trees can produce a large crop with only 10 to 20 percent of the flowers setting fruit, blueberry bushes need about an 80 percent set rate to provide an abundant harvest. The fact that blueberry flowers are receptive for only four to five days can often work against a high set rate. If the weather is cool, rainy or windy during that crucial time, honeybees won't be able to pollinate the flowers. Bumblebees, however, can save the day. They will work flowers at lower temperatures than honeybees and can fly in stronger winds. In addition, they're not as finicky as honeybees. Honeybees find some blueberry varieties more attractive than others. Coville, Earliblue and Stanley are less attractive than June, Rancocas and Rubel, for example. Bumblebees don't seem to care which variety they visit, as long as there's nectar there for them.

Once the flowers have been pollinated, the petals and stamens drop off, followed by the styles. The berries then grow rapidly for about a month. During this early growth phase, you may expect them to reach full size. If so, you'll be disappointed by their meager proportions and then frustrated when their expansion seems to stop altogether. Don't despair—just have a little patience. This slow phase is followed by a burst of final growth and then a ripening period. The average time from pollination to ripe berries is 50 to 60 days.

WHAT TO DO TO ENCOURAGE GOOD FRUITING

The most important thing you can do is to plant more than one variety of blueberry. This way you'll be providing for cross-pollination, which will give you more and larger berries. Blueberries aren't as touchy when it comes to compatible pollinators as some other fruits are. Highbush varieties all bloom at about the same time, so any two varieties will do. However, don't expect a lowbush or rabbiteye type to pollinate a highbush. You need to provide two different varieties within the *same type.* (For more on cross-pollination of the other two types, see the sections on Rabbiteye Blueberries and Lowbush Blueberries later in this chapter.)

A steady supply of water is critical to blueberries, all the way from flowering through ripening. If they undergo water stress during blooming, few flowers will set. A heat wave that settles in during blossoming can also be a problem, causing the flowers to wither prematurely. Blueberry plants that don't get enough water while the fruit is growing and maturing will ultimately produce smaller berries than well-watered plants. Remember that blueberries have shallow, fibrous roots that will dry out before those of your other fruit crops.

You can affect the size of your blueberry harvest even before you set your plants in the garden. Smart variety selection plays a key role. Like many other fruits, blueberries are sensitive to daylength. The varieties Coville, Earliblue and Jersey will produce plenty of flowers when days are between 10 and 12 hours long. However, fewer flowers will develop during a period of days longer than 12 hours. If you're a northern gardener and you plant one of these varieties, chances are you'll have a skimpy harvest. Daylength is the culprit. Since days in the North can be considerably longer than 12 hours in late August when next season's flowers are being formed, the plants' natural response is to form fewer flowers. Don't overlook daylength requirements when you're selecting varieties.

PRACTICALLY FROST PROOF FLOWERS

Blueberries are a good crop to grow in areas that are often hit with late spring frosts, for the flowers are remarkably frost resistant. In one Michigan study, flowers that were fully opened and exposed to 23°F were not injured. Even when the temperature dropped close to 21°F, some flowers survived. The varieties Concord, Rancocas and Rubel all retained 75 percent or more of their flowers intact at 21°F, while Pemberton lost 52 percent and Wareham lost 74 percent.

Blueberry varieties listed as late are a wise choice for an especially frost-prone area. They bloom a little later than early types, giving them a greater chance of escaping spring frosts. If a hard frost does threaten when your plants are in bloom, drape them with a light blanket. Be careful not to break the brittle branches when you cover or uncover the bushes.

RIPENING AND HARVESTING

It's nice to know that blueberries don't all ripen at once—otherwise harvesttime would be a little hectic! Fortunately, the harvest is spread out over a two- to five-week period. Within a cluster, the middle berries ripen first and are the largest. The outer berries that develop later must compete with one another for nutrients, so they're smaller and take longer to ripen. Blueberries that ripen when temperatures are above 75°F mature faster than those grown at cool temperatures, but their flavor isn't as good. In the warmer parts of the country, the earliest berries ripen in early July, while in the Far North, picking doesn't begin until August.

You can't judge a ripe blueberry just by its color, unfortunately. Blueberries turn deep blue before they are completely ripe. You have to test a few at a time to see whether they are ready to pick. Ripe berries will come off the bush very easily. If there's any resistance to your gently pulling fingers, leave the berries alone for a few days before trying again. Unlike some berries, ripe blueberries will keep for a couple of days on the plants without any loss in quality.

The unique dark blue color of blueberries comes from pigments called anthocyanins. Your particular varieties may have more of a light blue color; this comes from a waxy covering that can be rubbed off, revealing the dark blue berry underneath. The waxy bloom protects the berries from spoiling, so don't make it a practice to rub it off.

In fact, you should do everything you can to keep this protective coating intact. Since the bloom comes off more easily when the berries are wet, pick only under dry conditions. When blueberries are hand-picked commercially, they're dropped right into the individual baskets they'll be sold in so that the waxy bloom is disturbed as little as possible. You would be wise to follow this same practice, unless you plan to eat or process the berries within three days. If you can, pick directly into the bowl or freezer container you'll be storing them in. Just pouring the berries from one container to another can remove a significant amount of the protective wax. If you're going to use the berries right away it won't hurt to pick into a large container. When Diane does this, she likes to tie the bucket to her waist so both her hands are free for picking. Also, the berries have a shorter distance to fall than if she set the bucket on the ground.

You should be out there picking your blueberries every two or three days, but you can leave them for five or more days if you have to. Don't let them go for too long, though. Since the ripe berries tend to fall off the plants, a strong wind can knock loose ripe berries and reduce your yield.

Blueberries that have their waxy bloom intact can be stored under the ideal conditions we describe in chapter 1, What to Do with Ripe Fruit, for up to two weeks. Berries that are very ripe are quite soft and don't store as well as berries that are slightly underripe. If you mix the soft and firm ones in the same basket, they'll only last a few days in storage. You'd be better off segregating them in their own containers and using the soft ones as soon as possible. There is one problem with refrigerator storage—the berries tend to develop a tough, sometimes rough-looking skin. The flavor stays the same, but the berries aren't as pleasant to eat out of hand as freshly picked berries are.

RABBITEYE BLUEBERRIES

Gardeners used to growing highbush blueberries and hikers who have spent time in the mountains arduously picking wild blueberries or huckleberries would be amazed by the rabbiteye blueberry. These bushes can reach over 30 feet in height and produce prodigious crops of tasty berries. The name rabbiteye derives from the color of the berries as they ripen. Immature berries are light green and turn a pinkish red, the color of a white rabbit's eye, before maturing into the light blue color of the ripe berry. For a rundown on the various varieties that are on the market, glance at the Guide to Blueberry Varieties at the end of the chapter.

PLANTING AND CARE

Rabbiteye blueberries can be grown in less organically rich soils than highbush varieties. Even so, these plants will outdo themselves when placed in soil that's nice and humusy. While rabbiteye blueberries do best in acid soil with a pH around 4.2, they can survive a pH as high as 6 when planted in a hole fortified with large amounts of pine sawdust or sphagnum peat moss. Rabbiteyes also survive drought better than highbush berries. Just because they can tolerate dry conditions doesn't

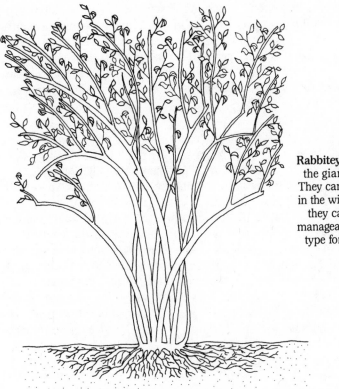

Rabbiteye Blueberry: These are the giants among blueberries. They can grow up to 30 feet tall in the wild, but in home gardens they can be pruned to a more manageable size. This is the best type for gardeners in the Deep South to grow.

mean you should skimp on watering, though. For the best harvest, plan on watering your rabbiteyes regularly, especially when berries are developing.

Even though they can withstand less than ideal conditions, giving these plants tender loving care really pays off. Studies done by a United States Department of Agriculture (USDA) scientist in 1980 showed that Tifblue rabbiteyes planted with about 10 quarts of peat moss added to the soil and watered adequately outyielded all the others. In fact, they produced three times as many berries as plants on which other cultivation methods were tried. Unlike most other fruit crops, rabbiteye bushes don't need to be fertilized. Fertilizer tends to make these vigorous bushes concentrate their efforts on getting even bigger, at the expense of producing bumper crops of berries.

Don't skimp on the amount of room you give rabbiteyes. Space the rows at least 10 feet apart, and separate the bushes by at least 6 feet. It's important to plant two or more different varieties of rabbiteyes to provide for cross-pollination.

If rabbiteyes have any drawback, it's their lack of hardiness. This characteristic really limits them to the warmer parts of the country. Dormant plants can survive temperatures of 18°F without damage, but they tend to break dormancy early, leaving themselves vulnerable to injury by spring frosts. Although highbush plants can survive temperatures down into the 20s even while in bloom, flowering rabbiteyes succumb at 30°F. Their winter chill requirement is much less than that of highbush varieties—just being kept at temperatures of 60°F during the day and 45°F at night for a mere 18 days will allow the plants to break dormancy. This dashes any hopes northern gardeners might have had that they could raise rabbiteyes successfully.

PRUNING RABBITEYE PLANTS

Dr. James M. Spiers of the USDA in Mississippi recommends starting right in with a pruning program. He advises cutting back rabbiteye bushes to a height of 4 to 6 inches and removing all weak shoots and fruit buds right after planting. For the first two to five years of the plant's life, he recommends a light annual pruning of all lower twiggy or weak growth, and dead or damaged shoots. This pruning can be done either in the dormant season or, with older plants, right after harvest. Once the plants are six years old, prune them just enough to keep the berries within reach and to prevent the plants from becoming too dense. Thin out older wood from the inside of the plant to open up the canes to sunlight.

Treat them right, and you'll see just how generous rabbiteye blueberries can be. While they take about four years to reach full production, once they get going you may end up harvesting nearly 20 pints of berries per plant over a 30-day season, starting around the end of May!

LOWBUSH BLUEBERRIES

Lowbush blueberries are only now becoming available to home gardeners (which explains why they're so hard to come by). Since these plants can be grown more successfully in some northern areas than highbush berries, their introduction into the home market is very welcome. The fact that they're super hardy doesn't affect their ability to produce abundant crops. A five-year-old planting, 18 feet long, can yield 16 pints of berries.

The key to lowbush hardiness is the way the plants grow. Unlike other familiar fruits, these plants propagate by way of underground spreading stems called rhizomes, much as quack grass does. The rhizomes form a branching mat 2 to 10 inches under the soil surface. This mat serves as "home base" for the roots that grow downward and the aerial shoots that grow up and out of the soil. Almost 85 percent of the plant's stem system consists of the rhizome mat. This distinctive system allows the plant to stockpile abundant carbohydrates and other nutrients safely underground for the winter, protected from the environment. Even if all the aboveground portions of the plant are killed by a severe winter, the rhizomes can send up new, healthy shoots in the springtime.

If you'd like to try your hand at growing these intriguing plants, the four named varieties (more accurately, clones) developed to date are Augusta, Blomidon, Brunswick and Chignecto. These clones yield from 40 to 60 percent more than many of the wild clones. Blomidon is the first variety actually bred for commercial production; all the others are clones selected from the wild. Turn to the Plant Sources Directory at the end of the book for the names of nurseries that carry these plants.

In order to allow for cross-pollination, you need to plant at least two different clones. As each clone grows and spreads, the clump will form a circle of new shoots, all of which are genetically identical. Year after year the clump will very gradually get larger. Just to give you some idea of how much these plants can eventually spread, some large ones in Canada, which are 150 years old, measure over 39 feet in diameter. Before you start to imagine your house being swallowed up by an encroaching horde of lowbush blueberries, you should realize that 150-year-old plants are a rarity, and the ones you'll be tending won't come anywhere near that size.

When you plant, set your bushes about 1 foot apart in rows spaced 3 feet apart. The pH of the soil should be anywhere between 4.2 and 5.2. If it is higher than 5.2, add sphagnum peat moss to the planting hole as described earlier under Planting Blueberries. Fertilize the bushes by mulching with peat moss or aged pine sawdust mixed with well-rotted manure. Water the plants when the soil dries out, making sure to water heavily so the moisture can reach down to the lowermost rhizomes.

Lowbush blueberries need sunlight to flourish and blossom. If they become overgrown or weedy, their berry production suffers. In nature, periodic fires burn and kill the plants that compete with the blueberries. The blueberry plants aren't

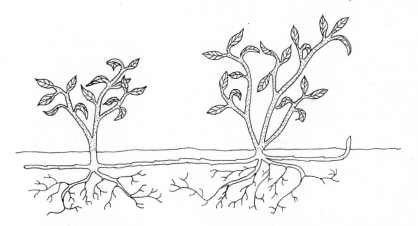

Lowbush Blueberry: This plant is small but will thrive in the Far North where other blueberries may fail. Lowbush plants spread by means of underground stems called rhizomes.

wiped out completely since their rhizomes are protected by the overlying soil from the heat of the fire. As long as the rhizomes are intact, the plants can always come back. Commercial growers also burn their fields to increase yields. Below the scorched earth the rhizomes quickly sprout new aerial branches, which form flower buds at the end of the summer and then blossom the next spring.

Growers who use this age-old method of pruning and weeding burn their plantings every two to three years. They do this in the springtime, before the plants have begun to grow, on a day when the soil is moist so the rhizomes don't get too hot. The first summer after burning, there's no harvest, but the following year's yield will be generous. Burning every two years actually results in a greater yield than burning every three years, so some growers establish two patches which they burn in alternate years; that way, they will get blueberries every year.

Large-scale growers burn their plants because it's a quick way to get the job of weeding and pruning done. For most home growers with modest-size patches, it's just as easy to hand pull the weeds and cut off half the aerial stems each year. The uncut stems will give you a crop that year, while the rhizomes from which you've cut the shoots will produce new ones for next year's crop. When the next season rolls around, you cut the stems that produced the previous year and pick berries from the year-old wood. With this method, half your plants will be producing heavily each year, and you never have to go a year without a crop.

PROBLEMS WITH BLUEBERRIES

One of the reasons blueberries are such a delight to grow is that diseases and insect pests are rarely a problem. Fruit flies sometimes infest the berries, but frequent applications of rotenone, starting when the berries set and continuing until the day before harvest, will keep them under control. (Be sure to rinse the berries thoroughly before you eat them.) Birds will almost certainly be the only real threat to your harvest. Fortunately, blueberry bushes are compact enough to make covering them with netting or cheesecloth relatively easy. Just be careful while putting on and removing the covering so that you don't break off any of the brittle branches. And be sure to cover the bushes completely! The birds are as anxious as you are to enjoy ripe blueberries—one small gap in the netting and they will beat you to the berries!

BLUEBERRY FRONTIERS

The future for homegrown blueberries is bright. The selection of improved varieties has barely begun with lowbush types, and the development of rabbiteye varieties is only in its infancy. Even highbush blueberries are still being improved for cultivation, with new and better varieties, often adapted to wider climatic variations, being introduced each year. Highbush breeders are concentrating especially on developing varieties that are hardier, produce blossoms with greater frost resistance, tolerate less acid soils, and ripen over a shorter period of time. Blueberry fans can dream of the day nursery catalogs announce a variety that combines all those characteristics, along with plumper, more flavorful berries!

Guide to Blueberry Varieties

HIGHBUSH VARIETIES

Berkeley—Extra large, light blue, sweet, midseason. Zones 5–8. Upright, spreading, very productive, medium hardy.

Bluecrop—Light blue, firm, somewhat tart, early. Zones 6–7. Upright, vigorous.

Blueray—Large, firm, sweet, early. Zones 5–8. Upright, spreading, medium hardy.

Dixie—Large, dark blue, late. Zones 5–8.

Earliblue—Large, light blue, very early. Zones 5–8. Upright, vigorous, hardy.

Herbert—Large, excellent flavor, midseason. Zones 5–8. Compact.

Ivanhoe—Large, excellent flavor, early. Zones 7–8. Upright, vigorous.

Jersey—Extra large, light blue, late. Zones 6–8. Large, spreading bush.

Northblue—Large, medium blue. Zones 3–7. Low, 20–25 inches tall.

Northland—Medium, very early. Zones 4–7. Low, spreading.

Northsky—Medium, light blue. Zones 3–7. Low, 10–18 inches tall, dense.

Stanley—Medium, midseason. Zones 5–8. Upright, vigorous.

RABBITEYE VARIETIES

Bluebelle—Extra large, midseason. Zones 7–9.

Briteblue—Extra large, light blue, late. Zones 7–9. Vigorous, moderately spreading.

Climax—Medium, early. Zones 7–9. Vigorous.

Delite—Medium to large, late. Zones 7–9. Upright, large, vigorous.

Homebell—Large, early. Zones 8–9. Upright, very vigorous.

Premiere—Large, very early. Zones 8–10. Upright, moderately spreading.

Southland—Medium, midseason. Zones 8–10. Upright, compact, vigorous.

Tifblue—Extra large, light blue, sweet, late midseason. Zones 7–9. 8–15 feet tall, vigorous. Resistant to heat and drought.

Woodard—Large, light blue, sweet, early. Zones (7) 8–9. Grows to 4 feet tall, many suckers.

CHAPTER 5

GRAPES
NATURE'S GENEROUS VINES

Vital Statistics

FAMILY	**HOW LONG TO BEARING**
Vitaceae	3–4 years
SPECIES	**CLIMATE RANGE**
Vitis labrusca (American)	American Zones 5–8 with some
V. rotundifolia (muscadine)	varieties in Zones 4 and 9
V. vinifera (European)	Muscadine Zones 7–9
	European Zone 8
POLLINATION NEEDS	
American, European and most	
muscadines are self-pollinating; some	
muscadines require cross-pollination.	

You don't have to think too hard to come up with a reason to grow grapes. Homegrown grapes can be used for juice, jelly, raisins and wine, as well as for fresh eating, and the vines make attractive additions to home landscaping. Across most of North America, grapevines can provide a delicious, abundant crop of fine fruit within two to four years. In addition to their rapid fruiting, grapevines have the advantage of long life.

Despite the rave review we give them, grapes aren't as widely planted by home gardeners as other small fruit crops such as strawberries and raspberries. There are several reasons for this. For starters, the price of one grapevine seems high at first glance. But when you realize that a single vine can produce many pounds of juicy grapes each year for 30 or more years, grapevines become a real bargain.

The processes of training and pruning grapevines may seem forbidding at first, but most gardeners find that, once clearly explained, they shrink back to the proportions of other, more common gardening tasks.

A VENERABLE AND VARIED CROP

Grapes were first cultivated by the Greeks around 4000 B.C., so today's vines have nearly 6,000 years of breeding and selection behind them. The European grape (*Vitis vinifera*), for example, has been developed into over 10,000 separate varieties! Because wine has played an important part in religious ceremonies as long as grapes have been cultivated, grapes have spread throughout the world with the expansion of religions. Christian missionaries, for instance, brought European wine grapes to the Americas very early in the settlement of these continents.

Native American grapes of a different species, *Vitis labrusca*, already grew here, and these were able to cross with the European grapes to produce new and interesting hybrids. These hybrids may show predominantly American or vinifera traits or a mix of the two. Add these to the many varieties of purely American grapes and all the European ones, and you can see that the grape is one of our most varied crops. Still another species of native grape, *V. rotundifolia* or the muscadine grape, is widely grown in the southeastern states and comes in many varieties as well.

THE GRAPE'S NUTRITIONAL PROFILE

Unlike most other fruits, grapes are not an especially nutritious food. They do contain small amounts of vitamins A and C, and raisins are a good source of iron. But grapes provide more sugar than anything else. The sugar is first produced in the leaves and is transported to the berries as they begin to color and ripen. When it enters the grapes, the sugar is in the form of sucrose, which is the same as table sugar. Within the grapes, the sucrose is broken down into the smaller sugar molecules glucose and fructose. Fructose, which is much sweeter than sucrose or glucose, gives grapes their wonderfully concentrated sweetness while allowing them to be relatively low in calories—only about 75 to 78 calories per quarter pound.

When you talk about grapes in their dried form, it's a different story. Like other dried fruits, raisins are very concentrated and thus calorie-rich; you get 164 calories in only 2 ounces (about 2 tablespoons). Golden raisins contain somewhat more sugar than dark ones. This difference is due to the way golden raisins get their lighter color. They are dipped into a hot chemical solution, bleached, and dried under cover. The solution kills the cells, thus stopping the metabolic breakdown of the sugars in the fruit. Darker, sundried raisins retain living cells longer and slowly consume some of their sugars as they dry.

HOW WILL YOU USE YOUR GRAPES?

You may be overwhelmed by the variety of grapes available, but one good way to narrow down your selection is to figure out how you want to use your grapes. There are grapes that are best for table use or fresh eating, others that make the finest raisins, and of course those that produce delicious wines. Sometimes a particular grape can be used for more than one of these purposes—these sorts of double-duty grapes are ideal for home growers who want versatility but don't have the space to grow lots of different grapes. Here we'll discuss the general types of vines that are available and how their fruits can be used. In the next section we'll discuss actual varieties in more detail.

Table grapes can be vinifera (such as Thompson seedless), American (Concord and Delaware), muscadine (Noble) or European/American hybrids (Cayuga). American grapes are preferred for juice and jelly since they stand up to the rigors of processing so well. When juice is canned, it must be heated to sterilize it, and wine grape juice develops an unpleasant "cooked" taste when it's sterilized. American grape juice retains its fresh flavor well under the same conditions. If you grow wine grapes and want to save some plain juice for drinking, there are ways you can preserve its fine flavor; don't use heat to extract the juice, and store it in the freezer.

American grapes are rich in pectin and make nice, firm jelly with nothing more than fruit and a light sweetener. Most European varieties have little pectin so you won't get a very good batch of jelly without adding commercial pectin. When canning grapes, use only seedless varieties, for the tannins (complex chemicals) in grape seeds become astringent when canned and can ruin the flavor of the grapes. If you use colored, seedless vinifera grapes like Beauty Seedless, some of the pigment from the skins will color the juice.

You may be surprised to learn that almost all the raisins made in the United States come from the familiar, pale green Thompson Seedless grapes. It's the drying process that changes their color. You can use any seedless grape variety to make raisins, but those with thin skins, a high sugar content and no seeds are best.

Although you can make wine out of any grape variety your heart desires, some make better wine than others. In general, wine grapes are either pure *Vitis vinifera* or hybrids with mostly vinifera traits. A good wine grape should be very juicy and have enough acid and tannin to give a wine character. If the grapes are too sweet, the resulting wine will be flat and almost sickeningly sweet. Grapes that make dry table wines usually don't make very good eating grapes because of their moderate sugar content (some exceptional hybrid varieties, such as Delaware and Himrod, are good all-around grapes). If you like sweet wines, you can grow Concord-type varieties and use them for fresh eating as well as for wine making. For a red wine with a deep, rich color, select varieties with small grapes. The color comes from the skin, and small fruits have more skin and less pulp than larger ones.

TIPS ON CHOOSING GRAPE VARIETIES

Although how you want to use the grapes may be uppermost in your mind, your choice also depends greatly on where you live. Perhaps more so than with other fruits, your success or failure in growing grapes will hinge on choosing the right type and varieties for your area. In general, American grapes prefer rather cool climates and can survive temperatures of −10°F or even lower.

Although they're hardy, American grapes can sometimes pose a challenge for northern growers. But if you really want to raise grapes, don't let conventional dictates about where grapes can and can't grow deter you. Both Dorothy and Diane, through careful choice of variety and vine location, have been successfully growing grapes in a part of Montana where winter temperatures can drop to −30°F and the frost-free season can be as short as 90 days. In some years it has been even shorter!

European vines require a frost-free growing season of at least 170 days (most need more than 180 days to mature a crop), and they won't survive temperatures below 10°F. Northern gardeners with short, cool summers might as well forget trying to grow vinifera grapes. Even if the vines survive, no crop will be able to develop and mature. It's a shame, but most of our continent just isn't suitable for growing the pure European wine grapes. Fortunately for wine lovers, there are many French-American hybrids, such as Aurora, which make fine dry wines and can be more widely grown.

Another problem with growing European wine grapes is their susceptibility to the deadly root aphid, *Phylloxera.* In the parts of the United States where this pest lives, raising a crop of European wine grapes is only possible by grafting vinifera stems onto roots of resistant native American grapes. If you live in such an area, you should be able to obtain grafted vines through local nurseries.

Potential grape growers who live in the hot and humid Deep South will also have little, if any, luck with European wine grapes. American grapes are likely to wither and die there, too, felled by the diseases that thrive in the South. (However, some hybrid varieties can be grown in the coastal southern states. Check with your local extension agent for a list of suitable varieties.) In warm, moist climates, muscadine grapes are the answer. Actually, they grow only in the South, for they require winter temperatures above 10°F and thrive in hot, humid weather. These vines can live up to one hundred years when well tended, while an American or European vine planted in the South would be unusual if it survived longer than six years. A long-lived vine like the muscadine will stand you in good stead, for this variety makes delectable jelly, juice and sweet wine.

For some help in choosing which American, wine or muscadine grapes to grow, turn to the Guide to Grape Varieties at the end of the chapter.

A CLOSEUP LOOK AT A GRAPEVINE

It's time to take a close look at a grapevine to see how it's put together. This sort of scrutiny will help you understand your own vines so you'll know how to get the most grapes with the least problems. If you examine a typical mature grapevine, you will see a trunk with shaggy bark growing up from the ground. While a bush or tree of a particular species has a characteristic shape, this isn't true of grapevines. They come in all sizes and shapes. A grapevine may have a very short trunk, close to the ground, or a trunk that is 12 feet tall! It may even have more than one. Branches may come off the trunk at different levels, or a cluster of branches may emerge from the top.

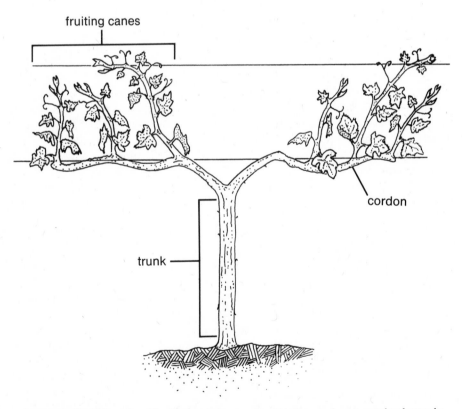

Anatomy of a Grapevine: The best way for you to get to know your grapevine better is to learn what the various parts are called. If you're well versed in grapevine anatomy, you'll find that following pruning and training directions is a breeze.

THE FOUNDATION

The shape of a grapevine is really dictated by the gardener, not by nature. The trunk of the grapevine is actually one of the first shoots, which was trained and pruned into its present state. If it doesn't have branches on its lower parts, that's because the buds and branches were pruned off as the trunk grew. This forced all of the growth to occur at the top, or head of the vine, the level at which the new shoots weren't removed. The branches that grow from the head of a grapevine are called arms. If there are branches emerging from the trunk that are trained to grow along wires, these are called cordons. Growing from the arms or cordons are spurs having shoots that flower and bear fruit; when these shoots turn brown and mature they are called canes. Near the base of the leaves on the shoots are small developing buds known as eyes. The following year, growth from the eyes will result in that year's fruiting shoots. The growing conditions of the fruiting shoots are very important, for the canes represent this year's crop and, by way of the eyes, next year's shoots and crop as well. As we go on to discuss pruning and training grapevines, it's important for you to keep in mind these different parts of the vine.

The leaves and tendrils of a growing grapevine have a definite pattern, depending on the species of vine. If you have a grapevine and don't know whether it's a European or an American type, look closely at the leaf growth pattern. *Vitis labrusca* varieties, such as Concord or Delaware, have either a tendril or a fruit cluster opposite every leaf on the upper part of the cane. On the other hand, muscadine and vinifera vines have a different pattern—a leaf-tendril or leaf-fruit cluster pair

tendril-leaf pair

fruit cluster-leaf pair

Telltale Pattern: An easy way to tell an American from a European grapevine is to examine the leaf growth pattern. The shoot shown here, with a single leaf between leaves that are paired with tendrils or fruit clusters, is from a European vine. On an American vine, a tendril or grape cluster appears opposite every leaf.

followed by a lone leaf. In all species, the leaves actually alternate up the shoot or cane.

The tendrils have an important job—they help the grapevine to support its often massive growth. The tip of the young tendril grows in a wide circle, away from the light, until it meets a support. When it touches something, the side opposite the point of contact grows rapidly, causing the tendril to coil tightly around the object. After the coiling is completed, the tendril becomes woody, providing a strong support for the growing vine. Diane found out just how tough these tendrils can become. One year she made the mistake of waiting until fall to remove grapevine tendrils that had grown between the boards of the wood siding on her house. What had been delicate green tendrils in spring when she first noticed them had metamorphosed into stubborn woody coils that were nearly impossible to pry loose.

WHERE FLOWERS FORM

In order to understand how grape flower clusters form and what determines their abundance, we must examine the developing eyes more closely. While they look superficially like single buds, each of the eyes actually contains three buds, any one of which can grow into a shoot. One of the three buds is larger than the others and is destined to become the fruiting shoot. However, if this main bud is destroyed or damaged, the other two buds can take over. The main vine (or the one that developed

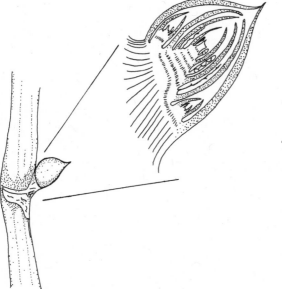

The Secrets Inside a Grape Eye: If you could look inside a grape eye, you would see three buds. The largest will become the fruiting shoot. In the event that this main bud is damaged, one of the other two can take its place.

from the largest bud) of all varieties can produce flower clusters, but, unfortunately, the secondary shoots in many varieties will not bear flowers.

Early in the growing season, while the current year's shoots are growing and developing, the vine is already preparing for the following year. You can't see them, but tiny embryonic shoots begin to develop within the eyes. The main bud can form about a dozen tiny leaves that are ready to burst out the next spring. Along with leaves, embryonic tendrils and flower clusters also form. How many of these flower clusters develop really depends on many environmental factors and is influenced by the hormonal system of the plant. A hormone called cytokinin is normally produced by tiny young rootlets in the spring and is transported up into the shoots. There, the cytokinin encourages the formation of embryonic flower clusters in the buds. It also serves as incentive for the flower clusters for the present year's crop to complete their development and blossom.

You may never have suspected that a plant's roots could directly affect the way it flowers, but that's certainly the case with grapes. If grape roots are subjected to poor growing conditions such as flooding or very cold weather followed by warm weather, you may face a dramatically reduced yield for two years in a row. Water-logged or chilled roots can't develop abundant cytokinin-producing rootlets, but warm weather still induces rapid vine growth. There's not enough cytokinin for the current year's developing flower buds or for the next year's embryonic flower clusters, so you're left with a meager harvest both years.

A bud's position on the shoot influences just how fruitful it will be. In many grapes, the first few base buds on the cane will produce little or no fruit. This is true of both Thompson Seedless and Concord grapes. Diane learned this the hard way one season when she pruned back her Van Buren (a Concord-type grape) vine to two buds. Each bud produced a healthy shoot, but neither had any fruit clusters, so she sadly had to do without grapes that year. Generally speaking, American grapevines are fruitful from the fourth bud from the base of the cane to the tip. Most European grapes have fruitful basal buds, but hybrids usually resemble their American ancestors when it comes to whether or not their basal buds bear fruit.

THE GRAPEVINE'S GROWING YEAR

Now that we've seen how a grapevine is put together, let's look at its annual growth cycle to learn more about how the vine's environment can affect the harvest. During the winter dormant period, the grapevine looks almost dead. But within its vine, the energy to bring it back to life in the springtime is stored in the form of carbohydrates. Other types of carbohydrates plug up the phloem, preventing any movement of nutrients through the plant. The stored carbohydrates also help lower the freezing point of the vine, protecting it from winter's cold. Grapes are especially vulnerable to winter damage, for during the dormant period the cambium is only one cell thick. If the weather is especially cold and the cambium is extensively damaged, the vines won't be able to produce new xylem or phloem tissue and may die.

THE VINE AWAKENS

As the temperature warms up in the spring, the phloem in an undamaged vine begins to unplug and the roots start to grow. The roots begin to transport hormones, nutrients and water back up into the vine through the xylem. Shortly after root growth begins, bud burst signals the end of the vine's dormancy. This occurs when the average daily temperature reaches around 48° to 50°F. At this stage, the growing vine and expanding leaves are using the food energy stored from the previous summer. During this period you can almost see the shoots grow—they can put on an inch a day when the temperatures are high. Just after bud burst, the miniature flower clusters that had developed the previous season also begin to enlarge.

The energy drain on the vine continues until the leaves are at least half grown. Once they reach that stage they can start to generate food for the rest of the vine through photosynthesis. Until the leaves take over, you can see how extremely important it is for the vine to have plenty of stored food to fuel rapid and vigorous early growth and good flower development. If you overprune your vine or let it bear too heavy a crop you're only dooming it to a slower, weaker start in the spring. When the carbohydrate reserves are perilously low, the flowers may simply not develop, and there won't be any fruit. Since the flower buds for next year begin to form shortly after this year's flowers get underway, depleted reserves can cheat you out of next year's harvest as well.

There are other factors that influence how fruitful the next year's buds will be. From mid-June to mid-July, the higher the temperature (barring excessive heat spells), the more flower clusters the tiny embryonic buds in the eye will develop. Temperature is so important to some varieties that no flower buds at all will form if the temperature isn't high enough! For example, the wine grape variety Muscat of Alexandria won't produce fruitful buds if the temperature stays at 68°F. Just four hours a day of 75°F temperature during the crucial period, however, is all it takes for the vines to form flower buds. You can see that if you live in a cool climate, varieties like Muscat of Alexandria will never give you a harvest. While almost all European wine grape varieties require high temperatures to initiate fruit clusters, a few (such as Riesling and Shiraz) will produce at 68°F. In general, American varieties will produce fruit at lower temperatures than European ones.

About six to eight weeks after bud burst, the vine's growth slows down and the flowers bloom. Now the older leaves are reaching maturity and can export lots of food first to the flowers and then to the developing fruit. The amount of nutrients available to the forming grapes influences the number of cells they develop—and this in turn determines their ultimate size. Vines overfertilized with nitrogen won't go through this important stage in which their growth slows down. Instead, the nutrients that should be feeding the young grapes will be channeled into greater vine growth. The fruit is likely to be aborted or at the very least the grapes will be puny and few in number. Since you want to eat grapes and not the viney growth, be sure to follow our guidelines in the section on Feeding Your Grapevines.

FROM FLOWERS TO GRAPES

A grape flower, compared with those of other familiar fruits, really isn't much to look at. Most fruits are pollinated by insects, and their flowers are designed to attract pollinators with colorful petals, sweet aromas and tasty nectar. (Just take a look back at the apple blossom in chapter 1 to see a prime example.) The grape, however, is self-pollinating, so it doesn't need to attract insects—which explains why its flowers are small and inconspicuous. (If you're a southern grower, you should know that many muscadine grapes can't self-pollinate. Some vines have only female flowers and no male flowers as a source of pollen. When you plant one of these varieties, intermingle vines with both male and female flowers with the all-female vines. When you buy your plants, the nursery will tell you what kind of flowers they have.)

Each cluster on a Concord vine has between 2 and 150 flowers, with 100 being about average. If you were to look at one of these tiny flowers more closely, you'd see that the petals are fused together into a tight little cap that covers the stamens and stigma. When the flower blooms, the petals detach at the base and split apart, and the entire cap pops off. This popping jiggles the thin, spidery stamens, releasing pollen onto the stigma of the flower.

Temperature regulates this interesting process of cap release; it must be at least 60°F or warmer before many flowers blow their tops. Rainy or cold weather can inhibit cap popping, which reduces the amount of fruit that sets. Even when the temperature is 60°F, there's no guarantee of good results. The grape flowers may

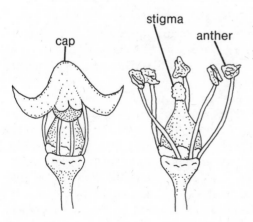

cap · stigma · anther

Pollination without Insects: Grape flowers can pollinate themselves just fine without any help from insects. On the left is the grape flower before pollination. The petals are fused together to form a cap that covers the stigma and anthers. When the flower blooms, the petals split open and the cap pops off. This shakes the anthers just enough to release pollen onto the stigma.

bloom just fine and have pollen deposited on their stigmas, but fruit set can still be low. The culprits are sluggish pollen tubes. At 60°F, the pollen tubes grow very slowly and don't reach the ovule in time to fertilize the egg. The difference temperature can make is really dramatic. At 60°F, it takes five to seven days for the pollen tube to reach the ovule, but at 90°F, it gets there in a few hours.

WHAT DO GRAPE SEEDS DO?

You may not appreciate grape seeds too much when you have to stop after every bite and spit them out, but you really should be grateful they're there. If they weren't, the grapes in your hand probably wouldn't be worth eating. The reason is that the number of seeds in the berry has a profound influence on its size. Grapes of seeded varieties contain from zero to four seeds each. Berries with no seeds or just one seed are very tiny and are called shot berries. Those with two to four seeds develop into good eating-quality grapes that vary in size; the two-seeded ones are always smaller than the four-seeded ones. Seeds produce hormones that encourage growth of the grape berry, and the more seeds, the more hormones.

What about seedless grapes? How do these manage to grow without any seed-supplied hormones? Commercially grown seedless varieties are sprayed with growth hormones to produce the big, juicy grapes you see in the store. But don't expect home-grown, unsprayed seedless grapes to come anywhere near their commercial cousins. Dorothy wasn't prepared for small-size grapes when she harvested from her seedless Himrod vines. While the fruit was abundant and delicious, the grapes were miniatures of the ones in the grocery store.

THE HOMESTRETCH TO HARVEST

Once the grape berry has set, it enters a phase of rapid cell division that lasts three to four weeks after blooming. Then the period of cell enlargement begins. At this stage, the berries are green and firm and full of acid rather than sugar—if you taste one, your mouth will pucker up. After five to seven weeks, the seeds begin to develop. This stage takes two to four weeks. Then begins the part the gardener waits for—the final ripening into sweet, tender, colorful grapes. All during this time you should make sure your vines get enough water. Water stress will cause berries to drop from the cluster and will make the grapes that hang on much smaller than normal.

The final part of the growth cycle is short, generally only a couple of weeks. All berries start out green, but at this stage they take on their characteristic color as they lose chlorophyll. (This means you shouldn't worry if you bought a blue variety and the grapes look green at first—they'll eventually turn the right color.) If they get lots of sunlight, their color will be that much deeper. Finally, once they've reached full size and have taken on their characteristic color, the grapes soften and accumulate sugar at a rapid pace. At this point, they're ready for harvest. The sugars come from carbohydrates stored in other parts of the vine which are transported into the grapes. If you've inadvertently overpruned the vine or if it's been shaded, fewer reserves will be available, and ripening will be slow. Once you've harvested the crop, the trunk and roots begin to accumulate carbohydrate reserves for early growth the next spring, and the whole cycle starts over again.

WHAT TO DO WITH A BUMPER CROP

Once a grapevine settles down and really starts producing, you may find yourself nearly snowed under by an avalanche of grapes. They don't keep for long in the refrigerator, but they can be frozen. When you thaw grapes, they will be soft and limp but their flavor will still be as rich and sweet as when you picked them off the vine. Juicing and drying are two other ways to preserve an abundant harvest.

Making raisins at home is easy, especially when you're growing seedless, thin-skinned varieties such as the European grape Thompson Seedless. If you are in the North you can use the hardy American varieties Himrod and Suffolk Red. Besides being easy, it gives you some control over what's in the raisins you feed your family. The raisins you purchase in the store may have been prepared using modern, "improved" methods which include placing the grapes on paper treated with malathion. This chemical not only kills insects attracted to the grapes during drying but also enters the raisins and helps kill insects during storage. Up to 12 parts per million of malathion are permitted in commercial raisins—but your own home-dried fruit won't have any.

The best grapes for drying are fully mature and very sweet. The sweeter the grapes, the plumper the raisins, because the fruit consists mainly of water and sugar. Once the water has evaporated, it's the sugar left behind that gives raisins their substance.

If you are a wine maker, you may have a hydrometer or hand refractometer for measuring the sugar in the juice of your grapes. Commercial raisin producers start to dry their grapes when they reach 17 percent soluble solids, but prefer to wait until they reach 20 percent, for at that level they will end up with heavier, sweeter raisins. At 17 percent soluble solids, it takes about 5 pounds of fresh Thompson Seedless grapes to produce 1 pound of raisins. At 20 percent, it takes around 4 pounds—significantly fewer grapes to get the same final result. Using the sweeter grapes also produces higher quality raisins—at 20 percent soluble solids, three-quarters of the final batch of raisins will rate as "high quality," while at 17 percent, less than half will rate that high. High-quality raisins have numerous fine wrinkles and are plump with a dark brown to bluish-brown color.

If you don't have a hydrometer, you can still dry your own raisins—just wait until the fruit has reached its peak of flavor but hasn't started to dry on the vine. Clip the bunches from the vines and remove any mashed or broken grapes. Cut very large bunches in half so they don't take too long to dry. Spread the grape clusters in a single layer in a shallow box or on a tray and set them where they'll get full sun. If your climate is very humid, you should bring the drying fruit indoors at night, and of course, if it starts to rain you should rescue your raisins-to-be.

In a week to ten days, the top grapes should turn brown. When this happens, turn the bunches over so the grapes on the underside will be exposed to the sun. The total drying time will be two weeks or more, depending on the weather and your climate. The raisins are ready when a bunch squeezed in your hand falls apart instead of sticking together in a ball. If you roll an individual raisin between your fingers, no juice should ooze out. As the raisins dry, most of them will fall off the stems.

Raisins can also be made in a food dryer, although it takes more work than sun-drying. You must carefully remove the grapes from the stems, without breaking their skins, and lay them out in a single layer. Drying them this way will take a few days. After the first 24 hours, check the raisins every few hours so you can catch them at just the right stage of dryness.

LOCATING YOUR VINES

Wherever you live and whatever sorts of grapes you decide to grow, you must give careful thought to locating the vines in your yard. Make sure they receive full sunlight and never plant them in a low-lying area that might be a frost pocket. Exposed flower clusters can be damaged by a temperature of 28°F, so it's important for you to protect the vines from late spring frosts.

There are several ways you can decrease the risk of frost damage. If your terrain is accommodating, plant the vines on top of a south-facing slope. Lacking such a slope, you can get the heads of your vines above frost level by growing them on high trellises. In New York State, grapes suffer less frost damage when trained to a 4-foot trellis than they do when grown on a 3-foot one. On the other hand, if it gets really cold (less than −18°F) where you live, you may want to consider doing the opposite—growing your vines on very low trellises. That way you can cover them completely during the winter with leaves or other loose mulch. (Don't take this protective covering off until you're fairly certain that the temperature won't fall below 25°F. If it should drop back down, be ready to run out and cover the plants again.)

Besides protecting your vines from winter cold and spring frosts, you need to make sure they get enough warmth during the growing season so the grapes can develop. In an area where summers can be cool, coddle your vines by growing them close to a south wall of your home or greenhouse. This microclimate will be considerably warmer than an exposed spot out in the yard and can make the difference between an abundant harvest and no grapes at all.

For the adventuresome grower who lives a little further north than most grapevines prefer, there's a way to pamper the vines and give them a head start. When you remove the protective mulch in the spring, cover the plants with clear plastic to form a mini-greenhouse. Be sure the plastic doesn't actually touch the vines or

Tactics for Short-Season Growers

Getting grapes to mature in short-season areas will never be fool-proof, but it can be done with careful planning. In the text we mentioned planting in the warmest, most protected part of your property and using a low trellis so you can cover the plants with a winter mulch. In addition, it's a good idea to maintain several potential replacement trunks, since winter damage is always a threat to your vines. In New York State, grape growers routinely grow new trunks every six years or so, because accumulated winter damage to the transport tissues can eventually inhibit the movement of nutrients and water from the roots to the shoots.

Never let your vines bear more grapes than they should. When too many clusters are left on the vine, ripening is slowed considerably. Grapes on vines with one cluster per shoot may ripen two full weeks before those on overcropped vines. You can also speed ripening by slowing the vegetative growth of the vines. When the grapes start to color, carefully cut back on watering. Just be sure the vine gets enough water to remain healthy. If the leaves wilt, that's a sign you're holding back a bit too much. Water your vine right away, but still try to water less often than before.

Your choice of varieties is crucial. Check catalogs carefully for the hardiest and earliest maturing varieties. Canadice, Himrod, Reliance, Van Buren, Worden and Ontario are all hardy, early varieties. As new varieties of grapes are released almost every year, the prospects for short-season gardeners are better all the time.

they may freeze if temperatures dip. Keep the plastic on day and night, venting it at midday to let excess heat escape. This warm, snug "cocoon" will encourage your vines to break dormancy and start growing sooner than they might if they were left to fend for themselves. Once the average daytime temperatures start to rise, be sure to remove the plastic completely—you don't want to cook your plants!

GRAPE ROOTS AND THE SOIL

Wine grapevines have incredibly far-reaching and vigorous root systems. In deep, finely textured soils, their roots will easily penetrate 12 feet down, and European wine grape roots can push as deep as 40 feet! A wine grapevine can fully utilize all the nutrients in 100 square feet of soil, its roots possibly stretching out as far as 8 feet from the base of the cane. This extensive root system allows a wine grape-vine to survive dry spells and to thrive in relatively infertile soil. As a matter of fact, it isn't advisable to give European wine grapes very rich soil. Fertile soil permits the vine to grow rapidly, and while this normally would be a good thing, for wine grapes it's a drawback. The crop will be large, but the grapes will be of poor quality for wine; they'll ripen more quickly than usual and won't have enough time to develop real character.

Even though American grapevines and wine grapveines grafted onto American roots have less vigorous and less extensive root systems than the European varieties, they don't require rich soil, either. For all grapevines, the depth of the topsoil seems to be more of a concern than an overabundance of nutrients. Experiments at the University of Arkansas Agricultural Experiment Station have shown that Concord grapevines grown in 20-inch-deep topsoil could produce almost 50 percent more grapes than plants grown in 10 inches of topsoil.

Grapevines of all types prefer well-drained soil. Older vines can survive flooded roots during the dormant period, but young vines may end up with fungus problems if they stay wet for too long. Remember when you fertilize and water your grapes that the feeder roots usually reach 3 to 8 feet out from the vine. So don't just apply water or spread mulch around the base of the plant—apply them in a nice generous ring to make sure all the feeder roots get their fair share.

SPACING THE VINES

Another noteworthy finding in the Arkansas Study is that the optimum spacing of vines differs depending on the soil depth. Vines in shallow soil gave their best yield when spaced 7 feet apart, while those in deep soil produced an even better crop when planted 10 feet apart. A good rule of thumb is, the more vigorous the vines, the farther apart they should be planted. Eight feet is a good distance for most varieties, but some European grapevines can be planted closer together to save space. Muscadines, on the other hand, sometimes require 20 feet for optimum growth. Check with your supplier when you buy your vines for the spacing recommended for your particular variety.

TEMPERATURE'S IMPORTANT, TOO

There's one final factor you need to consider about the soil your grapes are growing in. The temperature of the soil plays an important part in how well your vines grow. The main roots of Thompson Seedless grapevines, for example, need to be surrounded by soil that's 86°F for optimum growth. This is a degree of warmth that's almost impossible to achieve in a cool climate. Northern gardeners shouldn't despair—there are ways to increase the soil temperature around your grapevines. Bare, moist soil can be as much as 3°F warmer than ground blanketed with a low-growing cover crop or weeds. You can also pave the area around your vines with flat rocks to soak up the sun's heat during the day and slowly release it to the plants during the night. Or, you can mulch with plastic. Clear plastic will do a better job of heating the soil, but black plastic will keep weeds from proliferating. Whatever method you choose, be sure to warm the soil in an area large enough to cover the span of the wide-reaching roots.

FEEDING YOUR GRAPEVINES

Nitrogen is an important nutrient for good grape growth. American grapes and hybrids are especially sensitive to nitrogen deficiency, and they may produce a very disappointing crop if they're not getting enough. Like all good organic growers, you know that the key to avoiding a deficiency is to keep adding steady doses of a nutrient to your growing plants. With grapevines, you must time your application of nitrogen fertilizers just right. These plants need nitrogen most in the early spring and during their blooming period. Remember that each spring the roots develop new rootlets and root hairs that start the annual accumulation of nutrients. A good supply of nitrogen early in the season gets the vine off to a healthy start and begins the cycle that moves from flowering through fruit set and on to an abundant, sweet harvest.

It's very important to avoid giving nitrogen too late in the season or in too high a concentration, especially with winter-tender varieties. If you don't, the vines will grow rapidly, using up sugars that should have gone into the ripening grapes. This will slow down the ripening process and inhibit the proper coloring of the fruit. The vines will also continue growing too late into the fall and may be subjected to winter injury because the wood can't harden properly before cold weather sets in.

In chapter 1 under The Nutrients Fruiting Plants Need we describe the symptoms of a nitrogen deficiency. If you think your vines need nitrogen, add one of the nitrogen-rich organic fertilizers recommended there. Apply the fertilizer at least a month before leaf bud break in your area. (Grape leaf bud break occurs right along with apple and pear blossoming, about three weeks later than for peaches and plums.)

Grapevines can suffer from deficiencies of nutrients besides nitrogen. A lack of potassium will show up in the leaves, as we describe in chapter 1, and the fruit will be pale in color and may ripen unevenly. You should also keep your eyes open for signs of a zinc deficiency. The leaves will look puny and underdeveloped.

When your grapevines produce a lot of blossoms but don't set much fruit, your soil may need more boron. Grapes grown in the Southeast and some parts of the West are prone to develop boron deficiencies. The leaves of boron-deficient grapes are faded at the margins between the veins. If you don't correct the deficiency, the entire leaf except the veined area will pale.

Turn back to chapter 1 for pointers on which organic fertilizers will clear up deficiencies in potassium, zinc or boron.

TRELLISING YOUR VINES

Grapevines need direct sunlight and plenty of it. You can get at least twice as many grapes from a vine grown in full sun than from one that's shaded. Sunlight shining on the buds themselves is very important in encouraging fruitfulness the next year. Since the method of trellising you use affects the amount of sunlight that reaches the leaves and buds, you really should put some thought into choosing the best trellising method for your grapevines.

Your first inclination may be to run out and buy some grapevines, plant them, and then figure out how you're going to trellis them. Actually, you should figure out how you'll be trellising *before* you plant so that right from the beginning you can train the vines properly. If you install the posts and wires before planting the vines, their roots won't be disturbed in the process.

There are several things to consider when choosing a trellising method. How much space do you have? How severe is your winter climate? How many vines will you plant, and in what sort of arrangement? How vigorous are the varieties you have chosen? In this section, we give the dimensions for trellises used in commercial plantings in New York State. Shorter trellises can also be used. These are a help to short-season gardeners as we explain earlier under Locating Your Vines.

GENEVA DOUBLE CURTAIN

If you're planting several fast-growing vines of a variety such as Concord, the Geneva double curtain method of trellising is a good choice. This method opens the vigorous vines up to good air circulation and lots of sunlight, so there's less chance of disease developing. Since the plants receive so much sunlight, they're apt to give you ripe clusters of grapes a little earlier than usual. Vines trained to a double curtain are also unbelievably prolific.

The trellis consists of two T-shaped supports, which are 4 feet apart at the top of the T, and three wires. One wire is strung from post to post at a height of 4 feet. This wire will support the trunks. The other two wires are attached to the arms of the trellis supports; these wires will hold up the cordons.

Plant the vines in a row between the supporting posts (use the spacing recommendations made earlier under Grape Roots and the Soil). Prune each vine to two trunks that reach up to the trunk-support wire and secure them there loosely. Train the trunks to grow up alternate wires; that is, train every other vine to the front wire, and train the vines in between to the back wire. Allow each trunk to grow one 8-foot-long cordon. Take the two cordons from the separate trunks of one vine and train them in opposite directions along the same trellis wire. The new fruiting shoots will grow downward from the wires, forming two curtains of greenery, 4 feet apart.

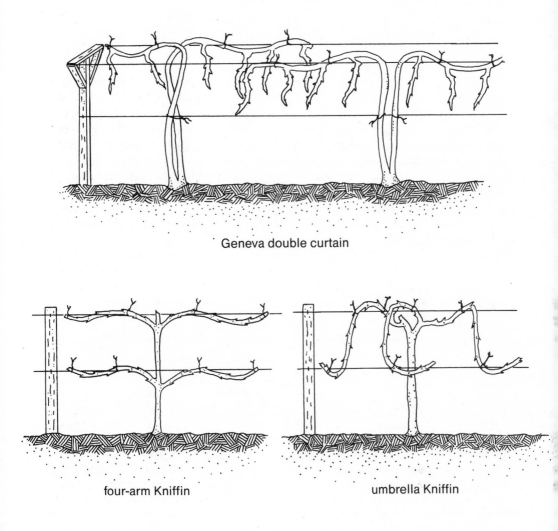

Geneva double curtain

four-arm Kniffin

umbrella Kniffin

Keuka high renewal

Gallery of Training Methods: Here are four of the most popular methods for training grapevines. The Geneva double curtain is good for vigorous vines and gives you a bountiful harvest. The four-arm Kniffin is an easy and attractive way to create a privacy screen. Vigorous vines do well with the umbrella Kniffin and produce in abundance. The Keuka high renewal is best if you're unsure about how fast and how large your vines will grow.

FOUR-ARM KNIFFIN

The four-arm Kniffin has been a popular trellising method for American grapevines because it's easy and attractive. This trellis is a good choice if you want to grow a leafy, "living" fence to border your property. But it does have a drawback. The lower parts of the vine are shaded, which can mean less growth and a smaller harvest.

The trellis consists of two wires strung between the posts, one at 3 feet and the other at 6 feet. With this method, you can use one or two trunks. With one, all four cordons grow from the same trunk; with two trunks, two cordons grow from each. No matter whether you go with one trunk or two, select a vigorous, straight-growing cane for each trunk. When it reaches the lower wire in the first year, tie it on loosely. When it reaches the upper wire, tie it there as well and prune it to that height. Allow four shoots (these will become the cordons) grow, two along the bottom wire and two along the top wire, and train them all away from the trunk. Keep each cordon pruned so that it's about 4 feet long.

UMBRELLA KNIFFIN

The umbrella Kniffin system uses the same sort of trellis as the four-arm Kniffin, but you train the vines differently. While a four-arm Kniffin vine is trained to four cordons, an umbrella Kniffin vine is head pruned (for more on this technique, see Training Head-Pruned Vines, later in the chapter). Remove all shoots as the trunk grows. Loosely tie the trunk to the lower wire and allow it to grow up to the top wire. A vine will usually reach the top wire in its second year of growth. Once it reaches that point, allow the head to develop 6 to 12 inches below the top wire. As shoots grow from the head, loop them over the top wire and let them grow downward. When they reach the lower wire, tie them to it. When you get done with all this, you'll see how this trellising system got its name—your vine will indeed resemble an umbrella!

There are some special advantages to a trellising system that involves bending the canes over the wires. For one thing, bending encourages uniform bud break, which will increase the crop. Also, the shoots that grow from the canes in the region before the bend are especially vigorous and are good to keep for next year's crop. Vigorous growers such as Catawba, Concord, Elvira, Fredonia and Niagara do especially well with this system.

KEUKA HIGH RENEWAL

With the modified Keuka high renewal method, you need to provide a trellis with 3 parallel wires spaced at 3 feet, 4½ feet and 6 feet. Like the umbrella Kniffin, this is a head-pruned system. You have your choice of growing one or two trunks on each vine. Allow the head to develop at the level of the second wire, and train cordons from the trunk along the lower wire. If you're growing vines that aren't particularly vigorous, you can keep the canes trained along the lower two wires.

With vigorous vines you can tie canes from the head to the top wire as well.

This system is useful if you're uncertain just how fast and large your vines will grow. It works well for varieties with an upright growth habit and a tendency to bear prolifically from base buds close to the trunk. The fruiting canes are shorter with this method than with the others we have described, so the yield won't be quite as high.

WHICH METHOD IS BEST FOR YOU?

All this information about trellising methods may sound very confusing; how do you choose the right method for your vines? In areas such as California, where warmth and sunlight are abundant during the growing season, the method of trellising is not of crucial importance. But in less than ideal climates, the trellising method

A Word about Trellis Posts and Wires

A good, sturdy trellis is vital to successful grape growing. If you throw together just any old assortment of junk you have lying around, the trellis may hold up for a year or two, but as the grapes get bigger

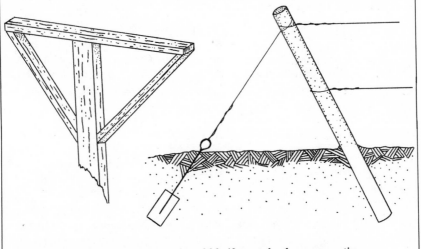

Extra Insurance: Your trellis won't wobble if you take these precautions. As shown on the right, set row-end posts in the ground at an angle. Imbed a wire tightener (available at hardware stores or make your own) into a block of wood and bury it in the soil to serve as an anchor. Run a support wire from the pole to the wire tightener. To stabilize a crossbar, add extra pieces of wood to act as braces, as shown on left.

(continued on next page)

A Word about Trellis Posts and Wires
Continued

and start to bear, it may not be sturdy enough to keep the vines and fruit aloft. Spend a little extra time and choose your materials carefully, and you'll have a trellis that can last for years. You can make reliable trellises out of wood, steel pipes or any other rigid material that can bear a heavy load. If you choose wood, consider investing in redwood which will resist rotting. Should redwood be too expensive or unavailable, you must treat your posts from top to bottom with a wood preservative. Choose one that contains copper naphthenate (Cuprinol is a common trade name); this is the safest of all preservatives and won't harm your plants.

For most trellises, you should start out with 8-foot-tall posts and try to bury them 2 feet (or as close to that as your soil will allow). That will leave you with a 6-foot stretch of post for your trellising wires. For extra stability, you should bury the end posts even deeper (3 feet is ideal), so you'll need to have at least two posts that are 9 feet long. Add an extra support to the end posts to make sure the wires are kept taut.

If you use a method such as the Geneva double curtain which has two high parallel wires, brace the crossarms with boards to form upside-down triangles. Use heavy, #11 galvanized wire to string your trellis. Wrap the ends of the wires securely around the posts, but use 1½-inch staples driven in with a staple gun to attach the wires to the crossarms. U-shaped nails such as those used to hold chicken wire will also work.

can make a great difference in the yield of the vines. In general, the result you want to achieve with trellising is to evenly distribute the foliage along the trellis, with a layer no more than two leaves deep. By avoiding a thick mat of leaves you're promoting good air circulation, an important factor in preventing fungus diseases. The fruit also will receive more sun, so it will develop a nice, rich color.

Here are some general rules that can help you choose. For vigorous varieties growing under optimum conditions, the Geneva double curtain, which reduces shading of the vines, is best. If you live in an area with harsh winters, you'll want to use a low trellis so that you can provide winter protection for your vines; the Geneva double curtain isn't appropriate for such locations.

Any variety with a drooping growth habit, large leaves and vigorous growth will do best when the renewal part of the vine (the canes or spurs on the head) is at the top wire. But less rapidly growing varieties with small leaves and upright development are most successfully trellised with a method such as the modified

Keuka high renewal, where the renewal area is located at the middle wire of a three-wire trellis.

Muscadine vines are so vigorous that they need plenty of growing room. If you decide to grow these vines on the Geneva double curtain, you can space them 8 feet apart. On an arbor, however, muscadines require 10 to 15 feet between plants. And if you decide to grow them on a fencelike trellis such as a four-arm Kniffen, with only two wires, you need to space them from 15 to 20 feet or more apart, depending on their vigor. With those muscadines that require pollinators, every third vine you plant should be a type with both male and female flowers.

PLANTING AND TRAINING THE YOUNG VINE

Now that you've chosen your grapes and decided on a trellising method, you can get down to the business of planting your vines and training them to grow the right way on the trellis. You should get your grapevines into the ground in the early spring, while they're still dormant (meaning the buds are not swollen), and once all chance of a hard frost is past. If your vines are already starting to break dormancy, you can plant them, but you must be ready to rush out and cover them at night if there's even the slightest chance of frost.

When you plant your vines, make the hole wide enough and deep enough so that the roots can spread out comfortably without having to be bent. You should spread them out at the same angle they were growing before. Position vines growing on their own roots so that only the first two buds are above the ground.

If you're planting grafted vines, keep the graft above soil level, otherwise it will produce unwanted suckers. These shoots that spring from the rootstock can be so greedy that they take most of the nourishment from the roots, leaving hardly any for the grafted top. When the suckers are especially vigorous, the top growth may even die.

Grafted vines that you purchase should have been disbudded, with all rootstock buds above the roots carefully removed. This preventive measure usually guarantees that you won't have sucker problems. Sometimes, however, the disbudding cuts don't run deep enough, and the vine produces a sucker anyway. If this happens, remove it right away so that no energy is drained from the grafted vine. Make sure you cut the shoot off flush with the surface of the rootstock to prevent the growth of new sprouts.

Cut back a plant that has more than two buds, or it won't have enough roots to support all the shoot growth. The first year your vine should be establishing a healthy root system and producing vigorous canes (one of these will become the all-important trunk). If you didn't manage to get your trellis in before planting, you can set up a temporary support good for the first year by hammering a stake in at the base of the vine. Tie the vine loosely to the stake to keep it growing upright. Before the second season rolls around, while the plants are still dormant, you should replace this stake with a full-fledged trellis.

If the vine doesn't reach the top trellis wire the first year, be ready to cut back to three or four buds at the end of the dormant season. It's best to wait until then so you can check for winter damage. Should the cane you chose as the trunk be discolored, or the buds be blackened, you can pick another cane that's in better condition to take over as the trunk.

Cut all canes but the future trunk off flush with the surface of the chosen cane. That keeps them from sprouting again. Prune back the trunk cane to three

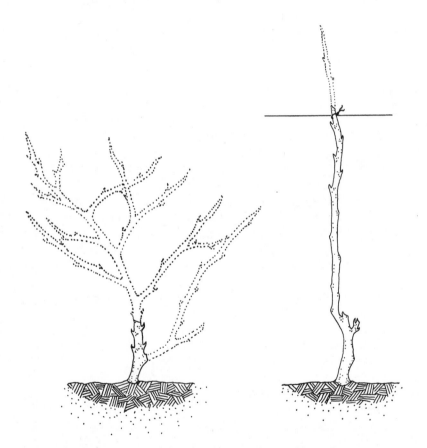

Pruning Pointers: On the left, you see the degree of pruning you need to do during the first dormant season. Prune back the trunk cane to three or four buds. As the shoots grow, select the most vigorous to train as the trunk and cut off all but one of the others. Prune the trunk renewal spur back to two buds. On the right, you see how the main trunk has reached beyond the trellis wire. Prune it at the level of the wire and tie the trunk to the trellis.

or four buds. When the new shoots are about 8 inches long, select the most vigorous one that's near the stake to be trained as the trunk and cut off all but one of the others; that extra one is your "insurance policy." Cut back this extra shoot to two buds so that it doesn't sap the main trunk cane's energy.

As the season progresses, rub off or prune away all the side branches that may grow off the main trunk. This forces growth upward. Don't remove any leaves you find growing on the trunk—they contribute food that fuels the plant's growth. Once the trunk cane has reached 12 to 18 inches above the top trellis wire, cut it off through the bud closest to the level of the wire. By cutting through the bud, you'll stop upward growth and end up with a healed-over knob that makes it easier to tie the vine to the wire—the tie will be less likely to slip off the top of the shoot.

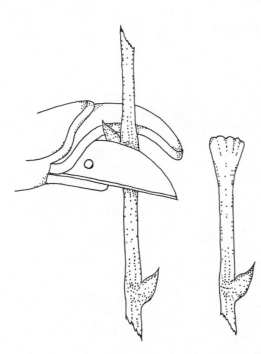

The Way to Cut: When you're cutting the top part of a vine to control its upward growth, always try to cut through a bud. The bud will form a swollen knob as it heals, giving you a handy place to wrap the wire around as you tie the vine to the trellis.

A NOTE ABOUT MUSCADINES

Muscadine grapes can be trained in whatever way appeals to you, but they must have the trellis in place the first year because of their vigorous growth. With these enthusiastic vines, most of the training will be completed the first year, and you can make the finishing touches the second year. By the third year, you'll be in the grape harvesting business with a good crop of fruit.

TRAINING THE CORDONS

If you've chosen a trellising method that uses cordons, after you've attended to all the initial pruning you'll want to encourage lateral growth along the wires. During the first season after planting (and sometimes into the second), remove any fruit clusters so that all the vine's energy can go into growing the cordons. When you're sizing up shoots to train into cordons, look for lateral shoots that are growing in opposite directions parallel to the wires. They should also be 6 to 10 inches below the support wires. Select two laterals below them to serve as renewal spurs in case the original cordons are ever damaged. Trim away all the other laterals. Let the shoots grow unfettered until they are 1½ to 2 feet long before tying them to the wire. This ensures that they won't break and that they will approach the wire at the correct angle. Tie them loosely, close to the point where they reach the wire. You should avoid tying them within a foot or so of the growing tip, or you may unwittingly

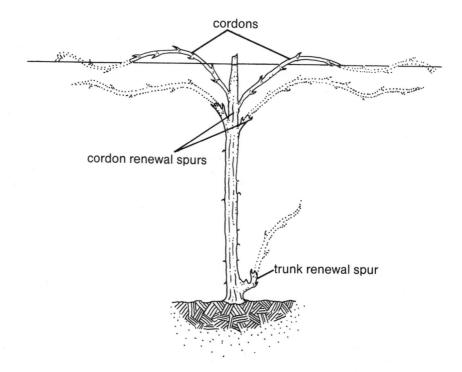

Cordon Training: Train two lateral shoots growing in opposite directions as your cordons. If you live in a climate where winters take their toll on grapevines, play it safe by keeping two cordon renewal spurs and a trunk renewal spur on reserve. The second season after planting, trim the cordons back to where the vine is ⅜ inch thick. Cordons that are too thin won't be able to support the weight of fruit-bearing shoots.

inhibit their growth. As they get longer, twist the cordons loosely over the wires to help hold them in place.

The next winter, trim the canes back to the point where the vines are at least 3/8 inch thick. If you allow a thin shoot to grow as the cordon, it won't be strong enough to support the weight of the shoots that will eventually grow from it. In addition to this structural weakness, the proper distribution of nutrients might be inhibited if the cordon became pinched against the wire. It may be necessary to cut all the way back to a point below the support wire before the vine is thick enough. If so, prune to a side bud that's just a few inches beyond the trunk. The bud's orientation will keep the cordon properly aligned with the wire.

If you're using the Geneva double curtain system, start with one trunk that gives rise to one cordon. The second year, let a sideshoot or sucker grow upward as the second trunk. From this trunk, you can train the second cordon along the wire in the opposite direction from the first one during the third season.

Whatever trellising system you use, always tie the cordons to the wires gently and loosely so they don't become pressed against the wire; remember, they will thicken each year as they grow older. You can use the same wire you used for the trellis to tie the vines; just to be safe, check the wires periodically to make sure they're not too tight.

As the vine buds come out in the spring, you must encourage the shoots to grow in the right direction. If you're using a trellising method with downward-growing fruiting canes, rub or cut off the buds along the top of the cordons. If your method calls for upright bearing canes, remove the lower buds.

You can equalize the growth of your cordons with a few snips of the pruning shears. Where one is growing faster than the other, cut back the more vigorous one. During the third growing season, you can harvest fruit from your vines if they are growing well. Leave just one fruit cluster per shoot, thinning all the others out. If your cordons haven't yet reached the length you want them to be, you're better off trimming away more of the fruit—leave one cluster on only every other shoot. This extra thinning allows more of the plant's energy to go into cordon growth.

TRAINING HEAD-PRUNED VINES

The early growth of a vine to be head pruned progresses just like one that's going to be cordon pruned. But after you've halted the upward growth of the vine by cutting through the bud closest to the wire, you're going to allow five approximately equally spaced lateral shoots to grow from the top of the trunk. Remove all the others and cut back the potential substitute trunk at the bottom of the vine. During the second growing season, you can let any vines that reached the top wire during the first season bear one bunch of grapes per shoot. But with slow-growing shoots, you can't afford to be as lenient—you must remove all the fruit so that the plants can devote their energies to vine growth.

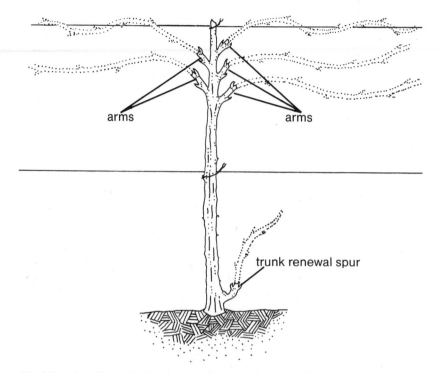

Head Training: Once the vine has reached the top trellis wire, select five lateral shoots to grow into fruit bearing arms and prune away all others. These five arms form the head of your vine. As they grow, twist them gently along the trellis wire. The trunk renewal spur is there in case disease or climatic extremes should damage the main trunk.

AN INTRODUCTION TO GRAPEVINE PRUNING

There's almost a bewildering number of ideas about which methods of pruning are best for grapevines. It's difficult to come right out and make a blanket statement like "This is the best way to prune grapevines." The three grape species and all their varieties differ so widely in terms of vigor and optimum growing conditions that it can be tricky to come up with pruning guidelines that encompass all their variations. What we'll give you in this section are the most useful general guidelines we've come across to help you with your pruning.

With grapevine pruning, as with so many other things in life, a happy medium is likely to be the best course to take. When a grapevine is left totally to its own devices and isn't pruned at all, too much fruit is produced for the size of the vine. This depletes the carbohydrate reserves. By the third season or so, the vine will be considerably weakened. It will be so low on reserves that instead of putting on a burst of early growth in the spring, the vine will only be able to manage slow, weak growth. As the season progresses, little or no carbohydrates will be available either for the current crop or for the next season's grapes. It's no wonder that unpruned vines eventually grow poorly and produce sparsely.

On the other hand, severe pruning can also weaken the vine. Because carbohydrates are stored within the dormant wood, you're cutting off food reserves with every snip of the pruning shears. You're also removing potential leaf area that would produce sugars for ripening fruit and carbohydrate reserves for the next winter. With many varieties, if you cut back so severely that only the first couple of buds are left, you've removed all the potential fruitwood from the vines!

BASIC PLAN OF ATTACK

The most reliable approach to pruning is to cut back the vines to manageable size so they don't overwhelm the trellis and to thin out the fruit clusters if the vine appears to be overproducing. (Signs of this are small berries that are slow to ripen, and a vine that puts on very little new vegetative growth.) Remember that the shoots need maximum exposure to sunlight, so choose the canes you remove with this in mind. Your goal in pruning is to open the vine up to the sunlight and to produce a good balance between vine and fruit growth.

When should you prune? The best time for pruning depends on your climate. In warmer areas, you can prune as soon as the leaves fall from the vine and it becomes dormant. But in colder regions it's better to wait until late winter, after the most severe weather has passed. Vines pruned in the fall won't be able to heal their wounds before the onset of cold weather. Pruning at the end of the dormant season rather than at the beginning also allows you to retain undamaged wood and cut off any canes that were killed by an exceptionally cold winter.

Young muscadine vines being trained to a trellis are the exception to these rules. You should prune muscadines during the summer. After selecting the vigorous canes that make up the vine framework, remove other growth so that the arms or cordons develop as quickly as possible.

BASIC TYPES OF PRUNING

There are two basic types of pruning—spur pruning, in which last year's bearing canes are cut back to the first two or three buds, and cane pruning, in which certain canes are selected for bearing and others are pruned out.

SPUR PRUNING

Spur pruning is the easiest method, but because the basal buds on American and most hybrid grapevines aren't fruitful, you can only use it on most European vines

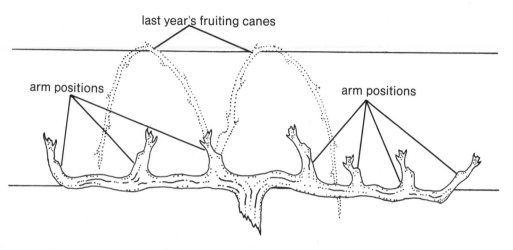

last year's fruiting canes

arm positions   arm positions

Spur Pruning: To set the stage, you need to establish a cordon-pruned vine, then select six to eight spurs that will serve as the sources for fruiting arms. During each season's pruning, you'll need to go back to each arm position and cut off one cane and prune the other back to two buds. From those buds will spring the following season's shoots.

and on a few hybrid varieties, such as Schuyler and Niabell. On a typical cordon-pruned vine, leave six to eight evenly spaced spurs, or arm positions, on each cordon the first bearing year. Each shoot will grow up to the top wire where you can tie it or loop it over. Then, when you prune the dormant vine, cut back each of those canes to two buds. The next spring, each bud will produce a bearing shoot. The following dormant season, when pruning time comes around, you'll find two shoots at each arm location. With varieties that produce large clusters of grapes, you should cut off one of the canes completely and cut the other back to two buds. This is the cane that will go on to produce a crop for you. If the vine produces small clusters of grapes and can, therefore, support a larger number of clusters, you can cut each of the canes back to two buds, leaving a pair of "rabbit ears" at each arm location.

Once your vines have reached this stage, all you have to do is prune each arm location back to two or four buds in the same way. Each year, the new shoots will arise a bit farther away from the cordon. An old spur-pruned vine will have large bumpy knobs, almost like medals of honor for a long life, marking each arm location.

CANE PRUNING

Cane pruning takes more thought and deliberation on your part than the more straightforward spur pruning. Each year at pruning time, you select bearing canes from the previous year's canes. Close to each cane you've chosen, you should also leave a two-bud spur. Each year the vine puts on two types of growth—new shoots from year-old wood which will flower and bear grapes, and new shoots from the

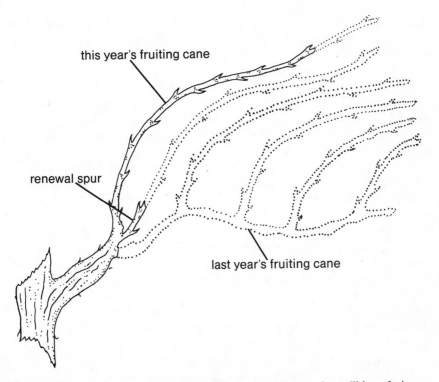

this year's fruiting cane

renewal spur

last year's fruiting cane

Cane Pruning: The rule to remember is that for each year-old cane that will bear fruit you need to leave a two bud renewal spur. The year-old wood will give you fruit that season, while the spur will provide canes for the following year's harvest. Here, last year's fruiting cane has been cut away. The two canes that developed from last year's renewal spur have also been pruned. The one that will bear the fruit has been cut back to 12 buds and the one destined to be the renewal spur has been cut back to two buds.

two bud spurs, which will usually not flower but will provide the canes for the following year's harvest. When you're choosing which canes to leave for fruiting and which to cut back as the renewal area, remember that canes exposed to the sun have the most stored carbohydrates. These will be the most vigorous growers. You can tell which they are by the color of the wood; sun canes will be darker than shaded canes, and the nodes will be closer together. Pruning to sun canes is always good advice. Cut back sun canes that will bear this season so they have from 8 to 15 nodes per cane. The high vigor Concord-type grapes do well with 15 nodes, while low vigor French-American hybrids produce best with 8 to 10 nodes.

THE KEYS TO BALANCED PRUNING

A critical question in both spur and cane pruning is how many nodes in total you should leave on the vine for fruiting. If you leave too many nodes and don't thin the resulting fruit clusters, you may weaken the vine so that it won't produce well in the future. On the other hand, if you prune too severely, shoot growth will be too vigorous. Not only will you get less fruit, but shoot growth will continue late in the season, resulting in a delayed harvest and immature wood that may not be hardy enough to last the winter.

You'll find that over the years, as you watch your vines grow and produce, you'll get a feeling for their bearing capacity. A mature, cane-pruned vine should have from 20 to 70 nodes left after pruning; 40 to 50 nodes is the happy medium that's best for most varieties. The minimum and maximum figures depend on the variety and how vigorous the vine is. (Vigor is something that's covered in the Guide to Grape Varieties at the end of the chapter.) The best procedure for the early years of growth, while you're getting to know your vines, is to prune moderately and then to thin the flower clusters so that each shoot carries only one cluster of grapes.

When you prune, first cut off all water sprouts—shoots that grow on wood more than two years old. You'd leave such shoots only if they were needed to replace a damaged trunk or arm. Next, cut off any winter-damaged wood. (This is the wood that's brownish-black and shriveled.) Finally, look over the canes and pick out the sun canes. Keep the thickest sun canes, and remember to cut a nearby cane back to two buds as a renewal spur. When in doubt about which canes to leave, choose those which will receive maximum sunlight and which are in a position to be tied easily to the support wires.

If you're not sure about how much wood to leave, prune less rather than more; you can always thin the flower clusters when they appear. And remember, the earlier you thin the grapes the better. With early thinning, you give the remaining clusters a chance to develop more berries. If you wait until the berries have set, it will be too late for any more to form. Be sure to use scissors to cut off the tiny clusters, or pinch them out with your fingers. Don't try to pull them off, or you might damage the shoot, which breaks easily at this stage.

Some varieties, such as Tokay and Malaga, also need to have the berries thinned in the cluster to avoid crowding. If the berries press lightly against one another, they will become deformed and a few may split and rot. To thin the berries, simply cut off the lower part of the cluster, leaving four to five branches laden with 80 to 100 berries. If you thin right after the fruit sets, the size of the individual berries will increase. Waiting even a week after fruit set, however, will lessen the size increase considerably.

PROPAGATING YOUR VINES

If you find that a particular vine does especially well in your garden, you can easily multiply it to give you more healthy, vigorous vines. Although most grapevines can be propagated by cuttings or layering, we're going to concentrate on the latter method, since it's the easier of the two. Also, since muscadines can't be multiplied by cuttings, we'll give the technique all home grape growers can use.

Time for Thinning: When your grapevine looks like it might overbear, come to its rescue by thinning flower clusters. On this shoot, pinching out the top cluster will encourage the remaining cluster to develop more berries. If you wait to thin until berries have already started to form, the clusters won't have a chance to set any more berries.

You don't have to be a master fruit grower to be able to layer a grapevine. In winter or early spring, choose a long cane that's growing in the direction in which you want to establish a new plant. Dig a long trench 8 to 20 inches deep, and place the cane in the trench so that only the top one or two buds are above ground. Before you bury the cane, wrap a wire snugly around it on the portion closer to the parent vine. As the new vine grows, the wire will slowly sever their connection. Since the phloem grows outside the xylem, the wire will prevent carbohydrates from moving from the new vine back to the parent vine, but the flow of water and minerals from the parent into the new vine will continue. Be sure to remove all leafy growth on the portion of cane that will be buried, and wait until the new plant is two years old before cutting it off from the parent vine.

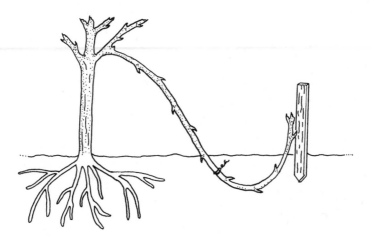

Layering a Grapevine: The easiest way to get two vines from one is to take a long cane, wrap a wire snuggly around it, and bury an 8- to 20-inch portion of the vine in a trench. As the new vine grows, the wire will slowly sever it from the parent plant.

PROBLEMS WITH GRAPES

Grapes have their fair share of pest and disease problems. The best way to keep the situation under control is to check your grapevines frequently. Remove any discolored grapes or leaves, and try to identify any pests before they become numerous and threaten to take over. Keep the area around your vines clean and discourage weeds; you'll deprive insects such as cutworms of a place to live and multiply.

The most serious problem for grapes is fungal disease. Various fungi attack different parts of the plant, from the leaves to the fruit. The best weapon against fungus is a healthy vine with open air circulation. If you have recurrent disease problems, you may find you need to prune and train your vines so that there are no overlapping leaves to provide a damp haven for the fungus.

Grapevines are hosts to many insects, most of which, like the Japanese beetle, are general garden pests. There are a few, such as the infamous *Phylloxera* aphid, which specialize in grapes. While this aphid can devastate a planting of European wine grapes in no time, its effects aren't usually as terrible on native vines and hybrids.

If you notice wartlike growths about the size of peas on the undersides of your grape leaves, your vine probably has a visiting contingent of *Phylloxera*. On sandy soils, the damage this insect can cause remains limited to the leaves. But on clay soils that crack when dry, the aphids can enter the cracks and attack the root hairs on the vine, stunting its growth. If you suspect an infestation of *Phylloxera*, there's really no cure. The only thing to do is to consult your local extension agent about resistant varieties for your area and replant.

Leaf hoppers—especially the variegated leaf hopper and the grape leaf hopper—can be a problem for grape growers. These are small insects, only about ⅛ inch long, but they can cause pretty large-scale damage if you don't catch them in time. The leaf hoppers live hidden on the undersides of the leaves where they suck the sap, causing whitish spots which later turn brown. If they damage enough leaves, your grape harvest can be affected.

You can get help from an unexpected source—a parasitic wasp that overwinters in blackberry plants attacks leaf hoppers. Early in the year the wasp feeds on another species of leaf hopper that is found on blackberry plants. Later on, the wasp searches farther afield and feasts away on any leaf hoppers among the grapes. If you've got the space, it's a good idea to grow blackberries near your grapes—the wasp can keep the leaf hopper population under control for you! Pyrethrum is also effective against these pests.

GRAPE FRONTIERS

Grape breeders are always looking for new vines with disease and pest resistance. They're coming up with interesting new hybrids that combine the fine qualities of vinifera grapes with the disease, cold and insect resistance of American species. Keep your eyes open as you peruse gardening magazines and nursery catalogs for the news that these new improved varieties are available to the home market.

Guide to Grape Varieties

AMERICAN VARIETIES

Alden—Blue, ripens very late. Zones 5–7. Moderately susceptible to downy mildew. Moderately vigorous vines. Use fresh, for juice, jelly, wine.

Buffalo—Blue-black, sweet, juicy, early. Zones 5–7. Resists downy mildew. Moderate to high vine vigor. Use fresh, for juice, jelly, wine.

Concord—Purple to black, aromatic, late. Zones 5–7(8). Resists downy mildew; susceptible to powdery mildew. Productive, high vigor. Use fresh, for juice, jelly, wine.

Daytona—Zones 8–10. Resists Pierce's Disease, good for Deep South.

(continued on next page)

Guide to Grape Varieties
Continued

AMERICAN VARIETIES

Delaware—Red, small, midseason. Zones 4–7. Moderately susceptible to downy mildew. Moderate to high vigor, good yielder. Use fresh, for juice, jelly, wine.

Golden Muscat—Gold, very late. Zones 5–8. Moderately susceptible to downy and powdery mildews. Moderate vigor. Use fresh, for wine.

Missblanc—White. Zones 7–9. Resists Pierce's Disease, good for South. High vigor. Use fresh.

Mississippi Blue—Blue. Zones 7–9. Resists Pierce's Disease, good for South. High vigor. Use fresh, for juice, jelly.

Ontario—White, early. Zones 4–7. Moderately susceptible to downy and powdery mildews. Moderate vigor. Use fresh, for juice, jelly.

Schuyler—Blue, early. Zones 4–7. Moderately susceptible to downy and powdery mildews. Use fresh, for juice, jelly.

Seneca—White, early. Zones 6–8. Very susceptible to powdery mildew; moderately susceptible to downy mildew. Moderate vigor. Use fresh, for juice, jelly, wine.

Steuben—Blue, sweet and spicy, late. Zones 5–7. Moderately resistant to downy and powdery mildews. Moderate vigor. Use fresh, for juice, jelly, wine.

Stover—White, early. Zones 7–9. Resists Pierce's Disease, good for South. Use fresh, for wine.

Van Buren—Blue-black, early. Zones (3)4–7. Moderately susceptible to downy mildew; moderately resistant to powdery mildew. High vigor. Use fresh, for juice, jelly.

Worden—Blue-black, sweet, midseason. Zones 4–7. Resists downy mildew; moderately susceptible to powdery mildew. Moderate vigor. Use fresh, for juice, jelly.

SEEDLESS VARIETIES

Canadice—Red, excellent flavor, early. Zones 4–7. Susceptible to downy and powdery mildews. Moderate vigor. Use fresh.

Glenora—Black, excellent flavor, midseason. Zones 6–8. Very susceptible to powdery mildew; resists *Phylloxera*. Moderate vigor. Use fresh, for juice, jelly.

Himrod—Yellow-white, excellent flavor, early. Zones 5–7. Moderately susceptible to mildew. Moderate vigor. Use fresh, for juice, jelly, wine.

Interlaken—Yellow-white, early. Zones 6–8. Moderately susceptible to mildew. Moderate vigor. Use fresh.

Lakemont—White, late. Zone 6. Moderately resistant to powdery mildew; moderately susceptible to downy mildew. High vigor. Use fresh, for juice, jelly.

Reliance—Red, excellent flavor. Zones 4–8. Moderately susceptible to mildew. Moderate vigor. Use fresh.

Remaily—Gold, midseason. Zones (5)6–8. Moderately susceptible to mildew. Moderate vigor. Use fresh, for juice, jelly.

Romulus—White, excellent flavor, late. Zones 6–7. Moderately susceptible to mildew. High vigor. Use fresh, for juice, jelly.

Suffolk Red—Red, excellent flavor, early. Zones 6–8. Excellent for the eastern U.S. Moderately susceptible to mildew. High vigor. Use fresh, for juice, jelly.

Vanessa—Red, excellent flavor, midseason. Zones 5–7. Moderately susceptible to mildew. Moderate vigor. Use fresh, for juice, jelly.

Venus—Blue, very early. Zones 7–9. Moderately susceptible to mildew. Moderate vigor. Use fresh.

WINE VARIETIES

(Because pure vinifera grapes can be grown in only limited parts of North America, they are not included here. All varieties listed are American or French-American hybrids.)

Aurora—Gold, early. Zones 6–7. Moderate vigor, bears heavily. Also use fresh, for juice, jelly.

Boco Noir—Dark red, midseason. Zones 5–7. High vigor.

Catawba—Red, very late. Zones 5–8. Moderately susceptible to mildew. High vigor. Also use fresh, for juice, jelly.

Cayuga—White, excellent flavor, late. Zones 4–7. Moderately susceptible to mildew. High vigor. Also use fresh, for juice, jelly.

Foch—Black, early. Zones 4–7. Moderately susceptible to mildew. Moderate vigor.

Fredonia—Black, midseason. Zones 5–7(8). Susceptible to downy mildew; moderately susceptible to powdery mildew. High vigor. Also use fresh, for juice, jelly.

Niagara—White, late. Zones 5–7(8). Moderately susceptible to mildew. High vigor. Also use fresh, for juice, jelly.

Seyve—White, midseason. Zones 4–7. Moderately susceptible to mildew. High vigor. Also use fresh, for juice, jelly.

(continued on next page)

Guide to Grape Varieties
Continued

MUSCADINE VARIETIES, FEMALE

Fry—Gold, midseason. Zones 7–9. High vigor. Use fresh, for juice, jelly, wine.

Higgins—Bronze, midseason. Zones 7–9. High vigor. Use fresh.

Hunt—Black, excellent flavor, early. Zones 7–9. High vigor. Use for juice, jelly, wine.

Jumbo—Black, midseason to late. Zones 7–9. High vigor. Use fresh, for juice, jelly, wine.

Scuppernong—Bronze, early. Zones 7–9. High vigor. Use fresh, for juice, jelly, wine.

Summit—Bronze, excellent flavor, midseason. Zones 7–9. High vigor. Use fresh, for juice, jelly, wine.

MUSCADINE VARIETIES, POLLINATORS

Carlos—Bronze, early. Zones 7–9. High vigor. Use fresh, for juice, jelly, wine.

Cowart—Purple, excellent flavor, late. Zones 7–9. High vigor. Use for juice, jelly.

Dixie—Gold. Zones 7–9. High vigor.

Magnolia—Bronze, midseason. Zones 8–9. High vigor. Use fresh, for juice, jelly, wine.

Noble—Black, midseason. Zones 7–9. High vigor. Use fresh, for juice, jelly, wine.

Southland—Black, midseason. Zones 7–9. High vigor. Use fresh, for juice, jelly, wine.

Welder—Green, midseason. Zones 8–9. High vigor. Use fresh, for wine.

CHAPTER 6

AN ORCHARD IN YOUR BACKYARD

When you have an apple tree right outside your back door, there's no excuse for having to put up with mealy, flavorless apples. The same goes for peaches, apricots, cherries and all the other tree fruits—you don't have to endure the insipid, bruised and expensive versions you find in the supermarket. You can satisfy your hunger for fresh, truly tree-ripened fruit by planting a tree or two (or even more) of your own. Besides their delicious bounty, the trees' natural beauty is an asset to your landscape. Fruit trees don't take much more time and attention than commonly planted landscaping trees such as maples, yet they make more productive use of your land.

You may have been hesitant to plant fruit trees because what you've read about pruning has been confusing or outright contradictory or because you're afraid you won't be able to pick varieties that will thrive and produce reliably in your climate. We understand your fears—we've both suffered from these problems. Dorothy made the mistake of following the advice in one book not to prune any fruit trees until they started to bear. As a result, she was faced with a yard full of trees that were developing in ways that actually worked against healthy, fruitful growth. Both of us have made substantial investments in trees which have died and others which have failed to bear because they simply weren't adapted to our area. There's no need to waste time, energy and money on trees that won't give you what you want—delectable fresh fruit. Once you understand something about how fruit trees function, you should be able to choose, plant, prune and care for them with confidence.

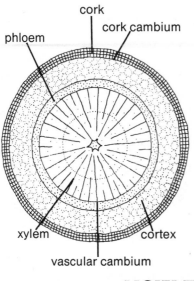

cork

cork cambium

phloem

xylem

cortex

vascular cambium

Peek Inside a Branch: If you were to look at a cross-section of a tree branch very closely, these are all the special structures you would see. Each part plays its own vital role in the life of the plant.

HOW TREES GROW

Trees grow in two ways. They add wood to their branches and trunks each year, increasing their girth, and they add length to the same areas by way of new, young shoot growth. To understand the yearly increase in wood, we must look inside a tree. The trunk, roots and branches all have basically the same structure. On the outside is a layer of dead outer bark, or cork. The outer bark protects the living parts of the tree from environmental dangers. Underneath the outer bark is a layer of growing, dividing cells which produce more protective, corky outer bark cells. This layer is called the cork cambium. The cork cambium is right under the bark on a young tree, but moves deeper into the trunk as the tree grows older.

Also underneath the outer bark of young branches and stems are various unspecialized cells that make up the cortex, a food storage area. The cortex disappears as the tree gets older. The most vital part of the trunk lies under the cortex. This is the area where the phloem and xylem are found. (See Anatomy of a Plant in chapter 1 for a rundown on the important job of transporting nutrients and water these tissues perform.) A ring of cells called the vascular cambium separates the phloem, which lies along the outside of the vascular cambium, from the xylem, which is inside. The vascular cambium produces replacement cells for both the xylem and the phloem. The whole region from the dead outer bark inward to and including the phloem is all considered to be bark. When a mouse or gopher eats all the bark around a trunk, girdling it, the gnawing destroys all the phloem and leaves some of the xylem intact. Unless you do some emergency grafting the tree will die because it can't transport water and nutrients past the girdled area.

Each spring when the tree starts to grow again, buds produce the hormone auxin, which activates the vascular cambium. The vascular cambium in turn generates new phloem and xylem cells. The old phloem cells are pushed outward by the new ones and eventually disappear within the cork layer. Since there are lots of buds producing auxin, much new growth occurs. Because the hormone is present in abundance, the first new xylem cells will be large. As growth slows down, less auxin is being produced, so the size of the xylem cells decreases. There's a distinct size difference between the smaller xylem cells produced late one season and the larger ones that develop the next spring. This distinction between late summer wood and early spring wood shows up as rings; when you chop down a tree or come across one that has fallen, you can tell how old it is by counting the rings in the cross-section of wood. Every year, the circumference of the vascular cambium increases as it rings the ever-increasing woody heart of the tree. As the tree gets older, the bark also gets thicker as new cork cells are added.

Sometimes, people will leave a low branch on a tree, expecting that as the tree grows, the branch will move farther off the ground. But once a branch is formed, it always stays at the same height, for the lengthening of the tree occurs at the growing tips.

If you look at a dormant fruit tree, you can see tightly closed, protected buds studding the branches. Some of these buds may be larger than the others—they are the flower buds. The smaller buds are vegetative ones that will sprout in the springtime to produce new branches. Most rose family trees bear their fruit buds on spurs which are actually compact branches with very short internodes.

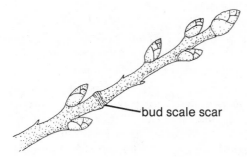

bud scale scar

Key to Measuring Tree Growth: The best way to tell how much new growth your tree has put on is to inspect several branches and measure from the tips to the bud scale scars.

MAN-MADE TREES

When you look at a cultivated fruit tree you see just one plant. But in reality, what you're looking at in most cases is two trees that have been combined together to make one. The roots or rootstock come from one tree, and the upper part of the tree, called the scion, comes from another. Sometimes, a third tree enters the picture, contributing a short section of trunk, called the interstem, between the roots and the scion. Why are parts of different trees combined? The reason is that some varieties of trees produce wonderful fruit and others have superior root systems. Rather than grow a tree on its own roots, which may have a serious problem such as disease susceptibility, an orchardist will graft the scionwood onto a rootstock variety that is known to be disease resistant. In this way grafting allows us to have the best of both worlds.

Rootstocks can offer other beneficial characteristics besides disease resistance. Many rootstocks dwarf the scion to varying degrees. That explains why we can buy Red Delicious apple trees that produce the same kind of fruit, but grow to different sizes. Some rootstocks can affect the way the scion bears fruit—speeding up the harvest, making it smaller, or making it larger. Still other rootstocks are especially efficient at absorbing particular nutrients and may be useful in areas with certain soil deficiencies. Along the same lines, some rootstocks are especially adapted to wet or sandy soils. Scientists really aren't sure just why the rootstock affects the scion, but whatever the reason, the results are certainly useful.

As you can see, your chances of success in fruit growing are often determined by the characteristics of the rootstock. It's unfortunate, but catalogs and garden centers usually don't let you know which rootstocks are used, though this is just the sort of information you need to select a tree that's suited to your particular climate and soil conditions. When you're shopping for trees and you don't find any information on rootstocks, ask about them. Write to the nursery or ask the manager of the garden center. Any clues you can get will help you make the best choice. (We give characteristics of certain rootstocks in the chapters that follow.)

INTERSTEMS TO THE RESCUE

Interstems are used for two purposes. Sometimes a particular scion doesn't graft well onto a desirable rootstock. In that case, an interstem of a variety which is compatible with both the rootstock and the scion is inserted between the two to serve as a sort of bridge.

Interstems are also used when the grower wants the deep anchorage of a certain rootstock but desires a dwarf tree. Dwarfing interstems make it possible to choose the best rootstock for a particular soil type and still have small trees. It's interesting to note that the length of the dwarfing interstem can affect the ultimate size of the tree—just a ½- to 1-inch difference in length can result in a larger or smaller tree. The longer the interstem, the shorter the resulting tree. So, if you buy a tree with a dwarfing interstem from one nursery and another tree of the same variety from a second nursery that uses a different interstem length, your two trees may end up being different sizes.

Sometimes dwarfing interstems are used in tandem with hardy interstems to come up with a tree that's both dwarf and tough enough to stand up to winter's cold. When you count the rootstock and the scion, that means a single tree like this has four different component parts!

HOW A GRAFT WORKS

When a graft is made, two pieces of young wood similar in size from two different trees are attached together right after being cut. The cells that were cut through on each part of the graft die, and each half of the graft forms a layer of cells at the site of the wound, called a callus. The callus cells grow through the dead cells within only a few days and join together. Then, they start to develop into the various stem tissues. In a successful graft, the new xylem cells from one part join up with those from the other part, forming a continuous water and nutrient conducting system. The same thing happens with the phloem. Close contact between the graft halves is crucial to the success of this process. If the two layers of callus cells aren't pressing against one another, there will be no union of xylem and phloem.

Grafting isn't always successful, for some varieties don't graft well onto others. Scientists aren't sure just what causes this problem, but they do know that on one side or the other, cells die instead of forming a healing callus. In addition, a waxy layer may form, which separates one side of the graft from the other.

Often, different species can be grafted onto one another—plums can be matched up with peaches or apricots, for example. Even plants that aren't very closely related, like quinces and apples, may form successful grafts.

HOW A NURSERY GRAFTS A TREE

You may find yourself gasping at the price of a grafted fruit tree. Here you are, paying more than ten dollars for a skimpy, spindly young tree that doesn't look like much and won't bear fruit for a few years. But grafted trees take a great deal of time and effort to produce. For a nursery to come up with an ordinary grafted tree, it must grow two different varieties in quantity. For an interstem tree, it needs three varieties.

Rootstock trees can be grown in various ways. If a seedling rootstock is being used, the grower plants seeds and waits for the young trees to reach the right size for grafting. Seedling rootstocks by their very nature are variable and the grower can't be sure just what their characteristics will be. Individual plants of a clonal rootstock, however, are genetically identical. They are propagated in ways that maintain the same genetic material from one rootstock to another, within the same type. The grower can predict their characteristics with great accuracy. That's why, for example, when you buy an apple tree on an M9 rootstock, you know that the rootstock will produce a tree of the same size and pest-resistant characteristics as any other M9 rootstock.

There are two ways to produce clonal rootstocks. In one, the grower mounds sawdust around the base and lower branches of a year-old, dormant tree. During the growing season, the branches of the tree under the sawdust sprout roots. At the end of a year, while the plant is dormant, the tree is uncovered and the rooted branches are removed. Now the grower has several pieces of rootstock to use for grafting from a single tree. Another faster and cheaper method to get clonal root-stocks takes place in a laboratory. With the widespread use of tissue culture, which we discussed under New Plants from Old in chapter 1, grafted trees may become less expensive in the future.

Scion wood is usually taken from "mother" trees which are cut back very heavily each year so they produce very abundant, vigorous wood. If the trees pro-duced from these scions are to be sold as certified, disease-free stock, they are kept free of insects and nematodes which could transmit virus diseases. The trees are also kept from fruiting because of the potential danger from pollen-borne viruses.

Once the grower has the rootstock, the scion and, if needed, the interstem, they can be grafted together to form a new tree. This tree isn't ready for market yet— it must grow for at least a year before you'll get a chance to buy it.

DOING YOUR OWN GRAFTING

You may decide some day to try doing some grafting of your own. Perhaps you have a favorite old tree, maybe even one that's an heirloom variety, that's dying, and you want to keep it going. Or, a friend or neighbor may have a variety which you especially like or which does well in your area. In either of these cases, with a little know-how, you can graft a branch or two of a different variety onto an established tree.

Grafting can come to your rescue if you've made a mistake in choosing varieties and find that your trees just aren't producing fruit reliably. By using an established tree as a foundation, you can graft on one or more branches of different varieties that are known to be good bearers. Grafting can also come in handy if you have a small yard but you want to grow a type of fruit that requires cross-pollination. Instead of growing two or more trees, you can grow *one* tree and graft on branches of a compatible pollinator variety.

There are two basic kinds of grafting you can use for adding varieties onto a tree—ordinary grafting and budding. With ordinary grafting, you add the wood from one tree to that of another. With budding, you cut buds out of one tree and insert them into the branches of another. In both kinds of grafts, the outcome is the same. Once the graft takes, you'll have a branch of a different variety flowering and bearing fruit on the parent tree. Usually it's best to stick with one plant species when you're budding and grafting; don't try adding an apple branch to a plum tree or a pear branch to an apricot tree.

With any sort of budding or grafting, you goal is to get the cambium layers of the two parts to touch so they can grow together. If they aren't touching, your graft has very little chance of succeeding. Always use a very sharp, clean knife so the edges of your cuts will be smooth and can fit together as snugly as possible. It's important to bind the grafted or budded parts together tightly so that there's pressure forcing them into close contact. The pressure will encourage the callus cells to join together for a successful graft union.

Budding is much easier than ordinary grafting, and some fruits, such as cherries and peaches, don't graft readily but do take well to budding. For those of you who are feeling adventuresome or who want to use grafting to overcome certain dilemmas we've described above, we'll give a brief lesson on how to go about budding your trees. For a more in-depth guide to home grafting, we recommend the New York State Extension Bulletin 75, *Topworking and Bridge Grafting Fruit Trees,* by G. H. Oberly. (You can get this bulletin by writing to the Distribution Center, 7 Research Park, Cornell University, Ithaca, NY 14850; send 50 cents.)

T-BUDDING

If you want to add a new variety to a young tree, T-budding is probably the easiest way to do it. T-budding is done in mid to late summer, after the buds for next year have formed and become dormant, but while the bark on the tree can still be easily peeled back. The time when these new buds are at their plumpest and ready for budding varies greatly from one part of the country to another. You can begin budding in mid-June in the South but may have to wait until August in the North.

Cut a 12-inch-long budstick with six buds from new growth. Right after you cut the wood, snip off the leaves, leaving just the petioles (the little stems that attach the leaves to the branch). This ensures that the budstick won't lose moisture. If you can't do the budding right away, wrap the budstick in plastic and place it in an insulated container such as a Styrofoam cooler out of sunlight. If you have to

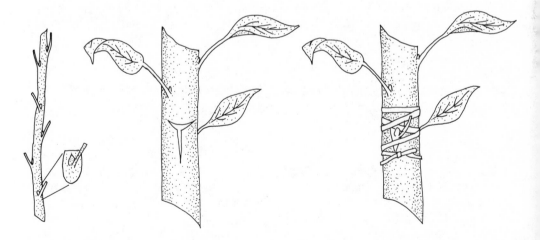

T-Budding Technique: This is the easiest form of grafting for home growers to master. Start out with a plump, healthy bud that you've carefully carved from a budstick. Slide this bud shield into the flaps of a T-shaped cut in the bark. Press the bark over the shield and bind the two tightly together.

leave the budstick overnight, put it in the refrigerator, with the lower end in a container of water.

When you're ready to bud, make a T-shaped cut on the branch where you want to add the bud. Choose a new branch that's about as big around as a pencil, and carefully slice just through the bark. Then pick out a nice fat bud from the middle of your budstick. Holding the budstick in your left hand, make a single smooth cut parallel with the wood, beginning ½ to ¾ inch below the bud and ending ½ inch above it. Slide the knife out. This cut should be just deep enough to remove a little wood. Press your left thumb on top of the bud shield to hold it in place (be careful not to damage the bud), and make a cut above the bud perpendicular to the first cut. This will free the bud shield from the stick. It takes a little practice to make these cuts correctly, but with six buds on the stick you have enough for a few trial cuts.

Slide the shield down into the flaps of the T-cut until the top of the shield is even with the top of the T-cut. If the shield is too long, carefully trim it until it's even. Make sure the bud is pointing upward, just as it was on the original tree.

Once it's in place, press the bark over the bud shield and fasten it securely using a long, thick rubber band that you've cut open. Wrap the band tightly, covering all the graft except the bud itself. Securely anchor the ends of the rubber band.

Leave the bud alone until next spring. When the bud has put on 2 to 3 inches of new growth—a sign that the graft was successful—you can cut back the branch to just above the bud. This pruning will force nutrients into the bud and encourage it to grow. Make the cut slant away from the bud. Don't leave too much wood above the bud, or it will just encourage decay.

In the South, where you can do your budding in June, the new bud may begin to grow that same season instead of waiting until the next spring. If this happens, cut back the growth above the bud to encourage it to grow quickly. This should help it to be in good shape for the winter.

PLANNING YOUR BACKYARD ORCHARD — A LONG-TERM INVESTMENT

The time you spend planning your orchard really pays off in the end. It's especially important to give plenty of thought both to the location you choose for your fruit trees and to the varieties you decide to buy. Not only are trees fairly expensive, but they also take some time to grow to bearing size. You want to make sure that you aren't wasting money or time when you plant them. Probably the two most common disappointments for home fruit growers are finding out that, for one reason or another, a fruit tree they worked hard to get into the ground really won't produce an abundant, ripe crop in their area and discovering that they haven't left enough space for their trees, which ultimately crowd one another and don't have enough room to grow properly.

Dream Trees—Beware before You Buy

You may have been intrigued by the descriptions in nursery catalogs of "5-in-1" or "4-in-1" trees, where several different scion varieties are grafted onto one trunk. Apples and pears are the most likely to receive this treatment. The catalogs make these trees sound like the answer to gardeners' dreams, especially if their yards are small. While these combination trees work out fine in some cases, in other case they don't live up to their promise. Unless the scion varieties are chosen with great care, various problems may result which cause nothing but headaches for the grower. The tree may eventually become a one-variety tree, defeating the whole purpose of its creation.

Let's take as an example a tree that might contain Red Delicious, Golden Delicious and Gravenstein scions. Golden Delicious grows slowly and bears early, while Red Delicious grows quickly and bears later. If you had such a tree, the Red Delicious part could quickly overtake the Golden Delicious, outcompeting it and overgrowing it. You would have to prune the Red Delicious with many thinning-out cuts to control it.

To further complicate matters, Red Delicious tends to grow in an upright fashion, while Gravenstein has naturally wider crotch angles. You would have to spread the Red Delicious branches at the same time you were pruning the Gravenstein branches to keep them from drooping. Without very consistent care, your tree could end up looking quite lopsided, and the more erect Red Delicious branches would outcompete the more horizontal Gravenstein ones.

Another factor to consider is that fruit varieties vary in their hardiness. A friend of ours here in Montana had a tree with Golden Delicious and Gravenstein scions. The Golden Delicious portion survived just fine, but the Gravenstein succumbed to an especially harsh winter, resulting in a tree that was dead on one side. When one of the varieties dies out like this, it can sometimes leave you with a pollination problem.

If you want to try a single tree with several varieties despite all the possible pitfalls, you should realize that it may need more attention than a regular tree to keep it in balance. In addition, you must carefully choose one with scions having similar growth habits, vigor and cold hardiness. In order for cross-pollination to occur, their blooming periods must also overlap; otherwise you won't get much of a harvest.

When you begin leafing through catalogs to find out what varieties and sizes of trees are available, you may get the wrong impression. Some catalogs give rather specific information that implies that a given type of tree—standard, semidwarf, dwarf—will grow to a particular height and width, regardless of the kind or variety of tree. This is simply not true. The ultimate size of a fruit tree depends on many factors—the kind of tree it is, plus the particular variety; the rootstock and scion used; and the climate, microclimate and soil type in your yard.

In each chapter on individual fruits, we give estimates as to how large the different trees may become. We say "estimates" because that is the best we can do. Take apples, for example. An apple tree described in a catalog as a semidwarf may be growing on any one of several different rootstocks, which will actually result in trees of various sizes. In addition, some scions, such as Gravenstein, naturally produce branches with very wide crotch angles, resulting in a tree with a large diameter. Other scions, such as some types of Delicious, have narrower crotch angles that give you a tree requiring a growing area of smaller diameter. Apple trees on dwarfing rootstocks can grow 1 to 2 feet taller than expected if you grow them in really deep, rich soil.

You'll even find that there's no such thing as a standard height for standard-size trees. These trees are generally grown on seedling rootstock, which is genetically variable. It's possible for two apple trees of the same variety grown on the same type of seedling rootstock to reach quite different heights.

THE ADVANTAGES OF SMALL TREES

Even if you've got the luxury of wide, open spaces, with plenty of room to plant all the trees your heart desires, you're still better off with small-size trees. With semidwarf and smaller trees you won't have to wait as long after planting for a harvest since they bear earlier than standard trees. They're easier to pick from, too—you don't need to perch on a tall, dizzying ladder to reach the fruit. Besides, the smaller trees will provide a more realistic-sized harvest for all but the few growers with a fixation on gallons and gallons of apple cider or a root cellar crammed with pears. And even though the trees themselves are smaller, the fruit they produce is full size. For all these reasons, we recommend that a home fruit grower choose a semidwarf or smaller tree if it's available in the desired variety. The only exception might be growers who remember a wonderful old apple tree from their childhood that was perfect for climbing and who are willing to wait many years before having a tree big enough to relive those memories!

We've mentioned that smaller trees come into bearing earlier, and you're probably curious to know just how much time they save. Semidwarf and smaller trees usually give you fruit two to four years after planting, while most standard trees often take much longer. For example, a dwarf apple tree should give you a good crop after just three years, while a standard apple tree can take up to ten years before producing enough fruit for the family.

Although they're smaller in stature, small trees are more efficient at gathering sunlight than their larger relatives. In order to produce fruit buds, the leaves of fruit-bearing wood must receive at least 30 percent of the available sunlight. This means that the leaves should be exposed to direct sun at least part of the day and to light shade the rest of the time. (Full sun all day long would be 100 percent.) A tall tree has many more leaves overhanging the lower branches than a short one, so less sunlight reaches the inner branches, no matter how well pruned the tree is. With an 8-foot tree, only 8 percent of the leaf area receives too little sunlight to develop fruit buds. With a 12- to 16-foot tree, that unproductive area increases to 13 to 19 percent. And for a 20-foot tree, the figure climbs to 24 percent. You can see from these percentages that you're better off planting several small trees instead of one large one—you will get a much greater yield per square foot of ground that way. Because of the difference in the amount of sunlight that reaches the fruit on the inner and outer branches of a big tree, the fruit will be more variable in size, quality and color than the fruit from a small tree that has received a more equalized dose of sunshine.

With small trees, you'll also find it easier to take advantage of the microclimates in your yard. You can grow the very smallest dwarfs in patio tubs, so that even a tree too tender to grow in your climate will thrive if you move it to a protected spot during the winter. Small trees can be planted closer to the protection of the south side of the house, too, expanding your fruit-growing possibilities if you live in a harsh climate.

YOU CAN'T IGNORE CLIMATE

When choosing trees for your yard, you must face up to the fact that different trees thrive in different climates. Apples grow best in areas with cold winters and are difficult to cultivate in the South. On the other hand, apricots and peaches are a challenge in the North. There are ways around these problems, as we've just seen, when you plant small trees. Selecting the right microclimate can be the key to success with a borderline fruit tree.

Once you've chosen what sorts of fruits to grow, you must select your varieties with care. Within one type of fruit tree you'll find an enormous range of varieties, each with its own unique set of characteristics. If you live in climate zones 6 to 8, you'll have a generous choice of fruit varieties, for most are well adapted to this region. Even so, you should check with your local extension agent for information about which varieties are most popular locally. Your area may have problems with particular diseases or pests, the humidity might be higher or lower than average, or the altitude may result in a shorter frost-free season. It's much easier to make allowances for these things while you're filling out the catalog order form than it is once you've got the tree in the ground.

A grower who lives south of zone 8 should choose low-chill varieties of fruits. (Turn back to chapter 1 under Dormancy and the Rest Period to brush up on what a plant's chill requirement means.) If you plant a tree that requires more chilling hours than your climate provides, the tree will be slow to leaf out, or it may not leaf out at all! A northern grower, on the other hand, must be sure to plant trees with high chill requirements. Low-chill varieties, if they aren't killed outright by winter's

The Name Game

Do you get confused looking through fruit tree catalogs and wonder what the terms dwarf, semidwarf, genetic dwarf, miniature, compact and spur all mean? Well, you aren't alone! These terms are, indeed, confusing, and no precise guidelines can be given as to their meaning. The terms you'll come across most often are dwarf and semidwarf. It's hard to give a specific set of measurements by which you can classify each. About all we can say is that when referring to a tree of the same fruit, a semidwarf is larger than a dwarf. Don't assume that a dwarf of one kind of tree will reach the same size as a dwarf of another—a dwarf peach, for example, gets to be 8 to 10 feet tall, while a "dwarf" cherry may grow to 14 feet in height.

There are two ways to get a small fruit tree. One is to cultivate a tree which by itself, without grafting, grows into a small tree. Such a tree is called a genetic dwarf. The other is to graft the tree, joining the top of one variety to a dwarfing rootstock.

A tree that's a genetic dwarf doesn't always go by that name in a nursery catalog. There are different terms used for genetic dwarfs of various types. A miniature tree is generally one which is small enough to grow in a tub; it only reaches 4 to 8 feet in height. For example, Stark Sweet Melody Miniature Nectarine is a genetic dwarf that grows only 6 to 8 feet tall and can be raised in a pot 18 to 24 inches in diameter. A tree that is listed as compact is one with internodes closer together than an ordinary tree, so it doesn't grow as large. Spur trees, developed from limbs on certain trees discovered to have shortened internodes, bear buds and spurs closer together than normal. These trees are also noteworthy because they produce more spurs and fewer vegetative

cold, may wake up too early in the North, making them vulnerable to late winter cold spells and early spring freezes.

Another important consideration is the rootstock. Just like scions, some rootstocks are hardier than others. A northern grower who chooses the hardy peach Reliance may still end up with a dead tree if the nursery grafted the Reliance scion onto a non-hardy rootstock such as Nemaguard.

Last but not least, if you live in a high-altitude or northern area, you must be sure to choose varieties that will bloom at just the right time to avoid frosts at both ends of the season. Apricots, cherries and peaches are most likely to suffer from late spring frosts, while apples present their biggest problem in the fall—ensuring that they ripen before autumn frosts hit. Although nursery catalogs often do point out which varieties have the hardiest blossoms and which bloom latest, they usually do not make clear which varieties require especially long ripening seasons. If enough home fruit growers were to write in requesting that information, perhaps nurseries would start to mention it in their catalogs.

Take Your Pick: When you're shopping for trees, you have your choice of miniature, dwarf, semi-dwarf and standard-size trees (shown from left to right).

branches than standard ones. Spur-type trees grow to only two-thirds the size of standard trees, so they are easier to pick from and to prune. Because they have such an abundance of spurs, these trees are very prolific.

When grafting is used to produce dwarfs, the results can vary in several ways. For example, a dwarf of a particular variety from one nursery may be grafted onto a different rootstock than the same variety from another nursery. Even though these trees bear the same name they may grow to different heights. In addition, a spur-type scion grafted onto a dwarfing rootstock will give you a smaller tree than will a standard scion grafted onto the same rootstock.

In the individual fruit tree chapters that follow, we give as much information about chilling requirements, earliness, hardiness and rootstocks as possible in order to help you pick the very best varieties for your yard. Be sure to check the Guides to Varieties for each type of tree fruit; they can help you sort out the sometimes confusing barrage of information all the different nursery catalogs give you.

SOME FACTS ON FRUITING

The fruits we discuss in the rest of the book belong to two categories. Apples and pears are pome fruits. The part we eat is derived from the fleshy outer tissue that surrounds the ovary. Despite all the research on fruits, scientists still can't agree on just which parts of the flower contribute to the pome fruit. All the other popular garden tree fruits are stone fruits, or drupes in scientific terminology. In these fruits the ovary wall has two distinct layers—the fleshy one we eat, and the other hard, stony layer that forms the pit's hard outer shell. The stone fruits almost always produce two eggs per flower, but only one develops into a fertilized seed. If you break open a peach or apricot pit, you can see the almondlike seed inside.

In general, pome fruits develop by a slow but steady increase in size while the growth of stone fruits occurs in three stages. The first starts logically enough with fertilization of the egg. Once that happens the ovary wall develops rapidly by cell division. This is the beginning of what will be the juicy fruit tissue. After the initial stage of rapid growth, a plateau is reached, and cell division activity slows way down. Then, as the ripening stage approaches, there is a dramatic increase in fruit size due to the expansion of the cells. The ultimate size of the fruit is determined during the early cell division stage, so it's important that your trees receive enough water and are otherwise kept healthy at this time. The first stage usually lasts from blooming until about one month afterwards.

Don't interpret tender loving care to mean massive quantities of nitrogen, though. If your fruit trees receive an excess amount, that first phase of rapid cell division will be prolonged. Too many cells will form, and later on in the fruit's development when those cells are supposed to be filling with flavor components, there won't be enough to go around. What you'll end up with is oversize fruit (a trait that tends to favor quick spoilage) that tastes watery.

PREMATURE FRUIT DROP

Almost always a tree will set much more fruit than will ultimately mature. Trees have a natural way of adjusting their load so they aren't completely exhausted by carrying too much fruit. They do this by dropping some of it at certain points in the season.

The reason behind this premature fruit drop has to do with seeds and a hormone they produce. This hormone stops the abscission layer from developing. Once the abscission layer appears, the fruit is cut off from the flow of nutrients and water from the tree. If the seeds in a particular piece of fruit don't get enough food because they're being outcompeted by the other fruit on the tree, they won't produce the hormone. An abscission layer will form, and the fruit will fall from the tree.

Fruit will also drop during periods when the hormone levels are naturally lower—right after the fruit begins to swell (in apples), and when the embryo in the seed is growing the fastest. This latter period is called the June drop and occurs in all the tree fruits we discuss, but it doesn't necessarily always happen in June.

BIENNIAL BEARING

Many fruit trees tend to develop a pattern of fruiting only every two years. This is called biennial or alternate-year bearing, and it can be very frustrating. One year you reap a bumper crop and have trouble finding uses for all your fruit, but the next year you harvest only a few pieces or nothing at all. A tree's ability to produce fruit is dependent upon its being able to generate enough energy to support vegetative growth, maturing fruit and future fruit bud formation. That's a pretty tall order, and if a tree is carrying a heavy crop, it simply doesn't have enough energy left over to set flower buds for the next year.

You can avoid this problem by properly thinning your fruit if the tree is over-bearing. (See the individual fruit tree chapters for specific thinning information; we also discussed thinning in chapter 1 under Miniature Food Factories.) The best way to thwart biennial bearing is to thin the fruits early, as soon as the first drop is over.

Alternate-year bearing usually becomes a problem when there is a severe winter or a late hard frost that kills the buds or blossoms. Because it isn't growing lots of fruit, the tree produces a flurry of flower buds that year, and the next year begins to overbear. If this happens to your trees, take action before the tree actually bears its bumper crop by thinning out immature fruit.

SPACING YOUR TREES

As we've already pointed out, it's difficult to predict just how tall a particular tree will get and how much space it will need at maturity. This doesn't mean that you have to guess blindly—there are reasonable guidelines you can follow to ensure that your trees have enough space. In each chapter, we give suggestions for planting distances for the various kinds of tree fruits. Remember that these aren't hard-and-fast rules and that there are various factors that can affect just how closely you follow them.

Ideally, you want your trees to be far enough apart so that they won't overlap their branches or shade one another when they're mature. On the other hand, you don't want them so far apart that they take up space you could use for other purposes. If you're planting a pollinator variety, you want to make sure the trees are close enough together so that bees can move easily from one tree to the other. For the best cross-pollination, make sure the pollinator tree is no more than three tree distances away (based on the planting distance for that particular tree). Sweet cherries are the only exception—the pollinator variety should be no more than two tree distances away. The type of soil you plant in also has a bearing on spacing. If your soil is deep and rich, your fruit trees need to be farther apart than in shallow or poor soil, since they'll grow more vigorously.

Climate also affects spacing. The experiences of apple growers in two different regions show just how much of an effect it has. After many commercial growers in Washington State successfully grew closely spaced apple trees, some Oregon growers in the Willamette Valley decided to follow suit. Unfortunately, their apple trees did poorly when they were in close quarters. If you analyze the two climates, you'll find that the Willamette Valley has many cloudy days and high humidity, while the Washington apple-growing areas are sunnier and drier. The Oregon growers found their trees didn't get enough sunlight with close spacing, and air circulation was poor. The decreased sunlight resulted in lower yields, and the reduced air circulation led to fungal disease problems. There's a lesson to be learned here—if your climate is humid and cloudy, you might want to give your trees an extra margin of growing room. They'll probably be more productive, and you're less likely to have to worry about fungal diseases.

PLANTING
IN LAWNS AND OTHER SITES

You may have a vision of a curtain of peach trees framing your front walkway, but you should be aware that there are special problems associated with planting fruit trees in a lawn. For one thing, you shouldn't allow the grass to grow right up to the trunk since grass roots will compete with the tree's roots for nutrients and water. In addition, you run the risk of damaging the trunk while cutting the grass. The best solution is to clear a circle about 6 feet in diameter around the tree. You can frame the area attractively with a circle of landscaping bricks set into the

Lawn Planting: When you plant a tree in the lawn, give it some breathing room. Never let the grass grow right up to the trunk. Clear a circle 6 feet in diameter and edge it with attractive landscaping bricks. Fill the circle with a layer of mulch.

ground. These bricks are L-shaped, with a flat skirt area. When you bury them so the skirt is level with the soil, the wheel of your lawn mower will ride right along the top of this skirt. This saves you from having to hand trim the grass. These bricks will also keep encroaching grass away from the tree. Metal or plastic edging material buried around the periphery of the circle will do the same thing. Top off the circle with a 3-inch-deep layer of mulch. A wide, mulched area around the tree is especially important for peaches and trees on dwarfing rootstocks, since their roots tend to be shallow.

When you plant a fruit tree in a lawn, you may have to alter your usual routine of lawn care to accommodate the needs of your tree. Heavy lawn watering during the fall can keep the tree growing actively when it should be slowing down in preparation for winter dormancy. Instead, do your heavy watering in late summer and then water the grass only lightly in the fall. In addition, you must be careful not to fertilize your lawn late in the season. Any nitrogen added after July in a cold-season area could stimulate the tree into late growth and keep it from going dormant in time to meet winter's cold. Fertilize the grass in spring instead, and you'll still have a green lawn without hampering your tree's growth.

If you're planting in a lawn where a previous gardener used fertilizers with added weed killers, or if you're siting your tree along a property line where a neighbor is likely to use these products, there are a few things you should be aware of. These products can kill young trees and may inhibit the growth or bearing of an older tree. The herbicide 2,4-D is especially harmful to fruit trees. If applied while the tree is leafing out, it may cause the leaves to turn brown and become twisted. Herbicides can seep way down into the soil and even damage the roots. In one study done in Washington State, herbicides were found as far down in the soil as 24 inches. Some were even discovered to have a residual effect. With these silent killers, by the time you see the damage, it is too late to do anything about it.

Since herbicides can be so devastating, you need to make sure they're not being used anywhere near your fruit trees. If your neighbor uses them, explain the facts to him or her and discuss your concerns. Perhaps you can persuade your neighbor to agree not to use them near your fruit trees.

VARIOUS WAYS TO PLANT A BACKYARD ORCHARD

Some home fruit growers like to set aside an orchard area on their property which they till to keep free of weeds. The trees, which are growing in bare ground, often do especially well in the early years. There are no weeds to compete with the young trees, and they grow fast. However, with this system, no organic material is returned to the soil. After a few years, the ground under the trees becomes compacted and shallow feeder roots may have trouble penetrating it. As the trees grow bigger, the area within the drip-line also increases. Since this area should be hand cultivated to avoid damaging the roots, there's more work for the orchardist to do with each passing year. Bare ground also tends to give up water faster than a mulched area, so it needs more frequent waterings.

By now you should realize that this planting scheme isn't the best for you or your trees. If you want to plant trees in a cleared, cultivated area, mulch them, don't leave them bare. Add a ring of mulch that extends at least 6 feet around each tree, just as you would for a lawn planting, and cultivate the unmulched areas to keep the soil loose.

Another alternative is to plant your fruit trees in an area that is otherwise left in natural condition. This method has its advantages and disadvantages, too. Planting fruit trees in a naturalized area can reduce your gardening work since you won't have to cultivate the ground between the trees. Natural planting can also reduce insect pests. Various sorts of parasitic wasps attack pests of fruit trees and use them to feed their offspring. The adult wasps feed on nectar from wildflowers. If you have clumps of wildflowers growing near your fruit trees, you're likely to have a bigger population of these helpful wasps in your orchard.

Now for the disadvantages: If this natural habitat is far away from the house and vegetable garden, your trees may not get as much attention as they need. Both of us have learned that it's all too easy to forget about trees that are out of sight, especially during the busiest part of the vegetable gardening season when there's so much other work to be done. Fruit trees need to be checked throughout the season to make sure they're growing well and aren't suffering from diseases, pests or nutritional deficiencies. During busy times, you may even forget to give your distant trees adequate water.

If you do decide to plant in a natural meadow, be sure to clear an area at least 6 feet in diameter around the trees so that they don't have to compete with the wild plants. Mulch the area to keep the plant growth down and to hold in moisture so you don't have to water as often. Be aware that deer and rodents such as mice and gophers are more likely to be a problem in a natural area than in a cultivated one. We give information on how to deal with these pests later in the chapter under Problems with Fruit Trees.

GETTING FRUIT TREES INTO THE GROUND

The ideal way to obtain a fruit tree is to buy one already planted in soil in a large container or to dig one up and immediately replant it in your yard. Unfortunately, most of us don't have access to trees like this. Instead, we must buy ours through the mail. You can order fruit trees in either the fall or the spring. In areas where the soil temperature stays above 50°F for several weeks after trees go dormant, fall planting is best. Any trees you plant in the fall, whether they come in the mail or from a local nursery, should be fully dormant. Remember, dormancy is the span of time from when a tree loses its leaves until it starts to leaf out again in spring. Although the tops of the trees are dormant, the newly planted roots will grow for awhile, getting better established before they are called upon to transport water and nutrients in the spring. If you make the mistake of planting a tree that isn't fully dormant and still has its leaves, it won't be as hardy during the winter as a fully dormant tree and may have problems making it through the cold weather.

Spring-planted trees should also be fully dormant when you set them into the ground so that they, too, can put down some new roots before the shoots burst forth. It's unfortunate for northern growers that some of the largest nurseries are in more southerly states; trees they sell can be on the verge of leafing out when they are shipped. If you happen to receive a tree in this condition, you should either return it or get it into the ground *fast.*

Sometimes, cost-conscious gardeners like to take advantage of inexpensive sale trees offered in late spring or early summer, after they have begun to leaf out. You should try to resist the impulse to buy such trees. Since replanting always results in some root damage, trees that have to bear the burden of developing leaves at the same time really end up struggling just to survive—and very often they don't make it.

WELCOMING THE NEW TREE

Mail-order fruit trees are sent bare rooted, usually with the roots wrapped in damp moss or wood chips to keep them from drying out. As soon as your tree arrives, unwrap it and check to see if the roots are still damp. If not, wet the moss or wood chips lightly. While you're at it, check to make sure the tree is alive. Slightly shiny bark means the tree is probably fine. If the bark looks very dark, dull or even black, rub off just a small bit of it near the top of the trunk to see whether the tree is green underneath—a sign that it's still alive. But if the area under the bark is dark brown or black, that's bad news. Check a little lower on the tree to make certain that it's dead. Instead of wasting your time planting a dead tree, ask for a refund.

If you don't have time to plant your tree right away, heel it in the ground in a shaded spot. See What to Do When Your Plants Arrive in chapter 1 for pointers. Anywhere from 1 to 24 hours before planting, immerse the roots in a pail of water. Be careful not break any of them while you're lowering the tree in. Don't soak the tree for more than 24 hours or you may suffocate the roots. This is especially important with apricots, cherries and peaches, all of which release hydrogen cyanide gas when the roots are unable to breathe.

Carry the tree outside in the water-filled bucket so the roots are kept moist right up until the moment the tree is placed in the ground. It takes less time than you may think for the fine roots, which are the most important ones for water uptake, to dry out and die in the sun.

PREPARING THE HOLE

While you're sweating away digging a nice, big hole, remember that the job you do now will affect your tree for years to come. If you have very rocky soil as we do, digging a good tree hole takes a lot of patience and determination, not to mention a strong back! Don't get impatient and quit after digging a small hole. It should be at least as wide as the natural spread of the roots, and you should be able to fan them out at the same angles they were growing before.

Plant your tree about an inch deeper than it was at the nursery, but make sure you don't bury the graft union. If you do, the scion may send out its own roots and bypass the desired effects of the grafted rootstock. For example, should you buy a tree on a dwarfing rootstock and plant it carelessly, the scion roots that sprout may cause the tree to grow to standard size! You'll recognize the graft union as a scar or slightly wider area on the trunk. The wood on one side may be a different color or texture than on the other side, too.

If your soil is very heavy, you'll want to make sure the drainage is sufficient before you plant. Dig your hole and fill it with water. If it doesn't drain out within 24 hours, you should try to find another place to grow your tree where the drainage is better. In the event that your whole yard is equally bad, you'll have to concentrate your efforts on improving the soil that will surround the roots. Add compost, peat moss or very well rotted sawdust to your soil in a one-to-one ratio. Coarse builder's sand will also enhance drainage. In very heavy soils, you may have to resort to drainage tiles. (Your extension agent can explain how to lay these tiles.)

If there are gophers or voles in your area, take protective measures while you're planting. That way you can avoid the awful surprise Diane had one spring morning when she looked out her window and saw her apple tree leaning to one side. When she went out to see what was wrong, she discovered that all the roots had been gnawed off, leaving a sharpened point at the end of the trunk. Dorothy, too, lost a peach tree to these underground marauders.

One way you can try to protect your tree from burrowing rodents is to line the sides of the planting hole with wire fencing with 1 inch holes. Make sure the wire extends from just above the soil surface down at least 1½ feet; 2 feet gives you even better protection. We have to add that this isn't a completely foolproof measure—some unfortunate fruit growers have reported encounters with intrepid rodents that seemed to be able to figure out ways to go around any sort of physical barrier that was set in their paths.

PLANTING THE TREE

When it's finally time to plant the tree, don't overlook the fact that this is the best time to position a supporting stake alongside. You'll be able to see where the roots are so you can nestle the stake in without injuring any. Too many people pound the stake in after they've planted—and they have no idea what sort of damage they're causing. A sturdy metal stake, 4 to 6 feet long, is best, since it won't rot. Place the stake in the planting hole 6 to 8 inches away from the trunk. When you're all done planting, tie the tree to the stake using a loop of soft cloth. The stake will protect the tree from wind stress and will stabilize it against encounters with careless children and dogs.

Once you've filled the hole, and the tree and stake are firmly in place, form a catch basin for water around the tree. About 18 inches away from the trunk, mound up the dirt a few inches high, forming a wall that will keep water from flowing away from the tree. It's a good idea to make another wall a couple of inches away from the trunk, forming a doughnut-shaped moat. The second wall will keep the water from pooling up against the trunk. This will discourage collar rot and will also keep water from freezing against the trunk and damaging it during the winter. When you're done with your moat building, add enough water to fill the moat to the top. This will give the tree a good long drink and will also help settle the soil around the roots and eliminate any remaining air pockets.

Now you need to protect the trunk from gnawing mice or gophers who relish young tree bark. The easiest protection is a spiral white plastic tree guard which you can buy from any nursery supplier. Wind the guard around the trunk of your tree (it should reach at least to the level of the first scaffold branches), and then just leave it in place. The space between the spirals will expand as the tree grows, so you don't have to worry about inflicting a slow death by strangulation.

If you're feeling industrious you can make your own tree guards out of ¼- to ½-inch mesh wire fencing. Cut a large piece with overlap to allow for expansion as the tree trunk grows. Wrap the first 18 inches of the trunk with the fencing and

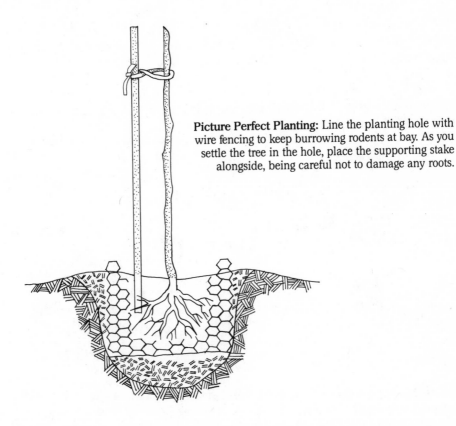

Picture Perfect Planting: Line the planting hole with wire fencing to keep burrowing rodents at bay. As you settle the tree in the hole, place the supporting stake alongside, being careful not to damage any roots.

bury the base 2 to 3 inches in the soil. Hold the fencing together at top and bottom with loosely twisted pieces of wire. Remember to adjust this homemade guard every year so that the wire doesn't girdle the trunk.

FOLLOW-UP CARE

There are a few more details you need to attend to before you can consider your planting job done. Because young trees don't have a full canopy of leaves to shade themselves and because their bark is thin, they're susceptible to sunburn. A spiral plastic tree guard can give protection, but if you use wire fencing, you should whitewash the trunk before you put on the tree guard. Be sure to apply the whitewash while the tree is dormant. You can use either commercial whitewash or white interior latex paint but never use an oil-based or an exterior latex paint. Oil-based paint can give off harmful fumes, and exterior latex may contain chemicals that can damage your tree. The white coatings will reflect the sun's rays and keep them from sunburning the trunk.

The whitewash will also protect the trunk from winter sunscald. On days when the sun is bright, the combination of sunshine and reflection of heat from the snow can warm the south side of the trunk until it's as much as 20°F higher than the shaded side. When night falls, the trunk temperature drops rapidly. The quick change in temperature can cause the bark to split, which may injure the cambium. The reflective coating will guard against this rapid rise and fall of temperature. (A white plastic tree guard will offer the same protection.) Many home fruit growers whitewash their trees for the first couple of springs until the bark thickens and is less susceptible to damage.

GENERAL TREE CARE

You might be tempted to think that once you've got your fruit tree in the ground, it can fend for itself. But fruit trees need their share of tender loving care to grow and produce a harvest. You wouldn't let your vegetable garden go for weeks or even seasons without watering and feeding it—and your fruit tree is no different. In this section, we'll give you some guidelines on how to tend your tree. In the chapters on individual fruits you'll find any specific instructions you may need to care for particular fruit trees.

MULCHING

No tree should be without mulch. A layer of compost or hay will conserve soil moisture and add nutrients to the soil as it breaks down. With a newly planted tree, wait until the soil has warmed up before you apply the mulch. With mature trees, be sure to renew the mulch every year so that it stays an even 3 to 6 inches deep. Keep the mulch away from the trunk to discourage any disease problems and rodent activity. Use the inner wall of the moat as the border for the mulch and extend it out 6 feet.

WATERING

Even with mulching, most fruit trees will need watering, especially if your climate is hot and dry during the growing season. Apple trees, for example, will absorb up to an inch of water a week. Soils that are abundantly rich in humus will store as much as 8 inches of water and give it up slowly to the trees, but most soils aren't anywhere near that efficient at retaining water.

With established trees, you should water once a week if it's hot and dry or once every two weeks if there's no rain but the temperatures are cool (in the 60s during the day). Young trees need even more water than older ones, since they're growing more rapidly and their root systems aren't as extensive. If you don't give your young trees enough water they will come into bearing more slowly than trees that receive an ample supply.

When you water, soak the ground thoroughly so that it's moist at least 12 inches down. Don't be one of those gardeners who goes out, hose in hand, and dribbles a little water on the ground for a couple of minutes. A light sprinkling does your trees no good—a thorough soaking takes time but it pays off in healthier trees.

If you've got only a few trees worked here and there into your landscaping, you'll probably find the garden hose the best way to water. But if you have quite a number of trees planted in roughly the same area, a drip or trickle irrigation system is likely to be the easiest and most economical. Drip irrigation beats out sprinkling on several counts. In a Washington State study, apples with trickle irrigation needed only one-third as much water as those that were sprinkled. Not only that, but the trickle-irrigated trees needed less pruning and spreading than the sprinkled trees. Trickle irrigation also keeps water off the leaves and fruit, so there'll be less chance of fungal problems developing.

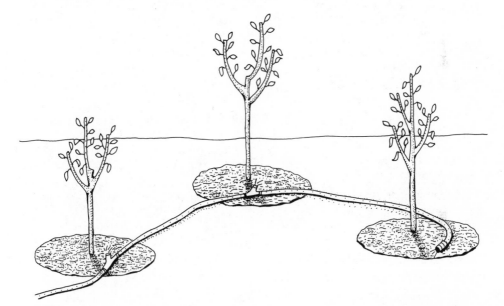

Trickle System for Trees: All you need to set up this system are some Y-connectors with shut-off valves from the local hardware store and ½-inch garden hose. Position the Y-connectors at least 6 inches away from the trunks.

Diane has set up a simple, effective trickle system that you might be interested in copying. At each tree, 6 to 12 inches away from the trunk, she places a Y-connector with two shut-off valves. The trees are linked to each other and to an outdoor spigot by inexpensive, ½-inch diameter hose. The branch of each connector that continues on to the next tree is left wide open, while the one to the tree is left open just enough to allow a trickle of water to flow out. The beauty of this setup is that it allows Diane to water all her trees at once. With this system you should leave the water on for two to four hours until the ground around the tree is very moist. To find out whether your tree has had enough, dig down 12 inches; if the soil is still dry, keep watering until the moisture reaches the 12-inch mark.

FERTILIZING

We have some very important words of warning about fertilizing fruit trees—never use manure or any other high-nitrogen material such as dried blood in the planting hole. If you do, you run the risk of having these materials burn the tender roots and overstimulate vegetative growth. Diane learned her lesson when she added rabbit manure in the planting hole for a cherry tree. At first, her tree seemed fine, but after three weeks or so, the leaves started to look burned, and as the season went on, the tree's growth appeared obviously stunted. With the wisdom of hindsight, Diane realized that the rabbit manure released too much nitrogen, which damaged the roots.

The best way to provide your young tree with a balanced diet is to give it a mulch of compost or hay fortified with manure. Add more mulch each spring and the tree should get just about everything it needs for good nutrition. Sometimes, though, a fortified mulch may not be enough of a nutrient boost. This is especially true if you've started out with a soil deficiency of some sort. In chapter 1, under The Nutrients Fruiting Plants Need, we try to paint a picture of what the various signs of nutrient stress are. A few fruit trees have their own set of symptoms and we mention those in the pertinent chapters. When we discuss individual fruit trees we also let you know how much growth to expect in a season so you can judge how vigorously your trees are growing. Watch them closely and keep an eye out for any indication that there's a problem. Pay particular attention when your trees are developing their crop—many are also forming buds for next year's fruit at the same time. This is a heavy energy load for trees to bear, and if they're not getting enough nutrients, deficiency symptoms are very likely to show up at that time. Remember, the earlier you detect a deficiency, the sooner you can correct it.

WINTER CARE

We told you in chapter 1 under Plants and Cold Weather that the leaves on a tree sense a change in daylength. As the days grow shorter in late summer, the tree begins to prepare itself for winter by forming a hormone that slows certain parts of the tree's growth. This same hormone also sets the stage for new growth next year. The hormone not only brings the growth of branch tips to a halt, but it also causes winter resting buds to form. These resting buds can contain tender, tiny flowers and leaves that will spring to life next season. They're encased in bud scales, which are specialized, thickened, waxy leaves that blanket the buds like a winter overcoat. The scales protect the buds from winter's drying winds and cold temperatures.

There are certain things you can do to help your tree prepare for winter—like cutting down on the frequency of watering in the fall and holding back on adding nitrogen fertilizers after early spring. But once winter arrives, there isn't much you can do for your tree, except hope that there aren't any early or late cold spells that could cause damage. If you've chosen your tree carefully and picked a variety that should do well in your area, chances are the tree will make it through the winter with flying colors.

In late winter, after the coldest weather is over but before the buds swell in anticipation of spring, it's time to venture outside to look over your tree to see whether it has been damaged. If you see any suspicious-looking shriveled areas, you can check for winter injury by gently rubbing off the bark from a small area to see whether the inner bark is green (just as you did when the tree first arrived). You may find that some twigs and branches have been killed; prune these off, cutting back to wood that shows signs of green. If winter damage has been severe, limit your pruning to the dead wood and don't try to shape your tree. A damaged tree needs all the leaves it can put out to recover.

Even a tree with severe winter injury can be saved by diligent watering. One of Dorothy's Golden Delicious trees is living proof of that. The poor tree lost two of its three main branches the first winter after it was planted. When the weather warmed, the remaining branch leafed out very weakly, indicating that the trunk itself had been harmed. Dorothy feared that the tree was lost and was ready to replace it, but Diane encouraged her to keep watering it to compensate for the damage the cold had caused to the conducting tissues. After several weeks of heavy watering, the tree slowly recovered and is now a lovely, graceful accent in Dorothy's small orchard.

If the trunk of your tree has become sunscalded and the bark has split, there is a type of grafting called bridge grafting that can save it. (See the earlier section, Doing Your Own Grafting, for the name of a booklet that can help you try your hand at bridge grafting.) If no more than one-quarter of the diameter of the trunk is affected, breathe a sign of relief. The tree should be able to recover on its own without grafting.

THE IMPORTANCE
OF PRUNING AND TRAINING

If you've done any reading on the subject, you've probably found that books on growing fruit trees devote a great deal of space to discussing techniques for pruning and training. But the advice is often unclear, and the reasons for using particular methods are rarely, if ever, explained. Like us, you may have been tempted to throw your hands up in despair and just not bother to prune your trees. But if you go to the trouble of buying and planting fruit trees, you might as well get the biggest crop of high-quality fruit those trees can give you—and that is only possible when you prune and train them properly. Understanding why certain techniques are preferable with a particular type of tree will make it that much easier for you to know what to do when faced with a tree in need of pruning. Because we believe so strongly that small trees are best for backyard growers, our pruning and training instructions are geared primarily toward small-size trees (semidwarfs and smaller).

Fruit trees have specialized areas where they form flower buds. Apple trees, for example, have tiny branches called spurs, which carry the flowers. Other trees, such as peaches, bear most of their flowers on the previous year's growth. If the spurs bear for many years, as sweet cherries do, you won't have to renew the fruit-bearing wood often. On the other hand, if the spurs are short-lived, as on sour cherries, you must prune to encourage lots of new wood with new spurs. In each fruit tree chapter, we let you know where flowers are formed and how long the fruiting wood keeps bearing so you can plan your pruning accordingly.

All fruit trees need to have plenty of sunlight so they can produce an abundance of food. They need this energy for growing new wood and developing flowers and fruit. They also need an extra stockpile of food for winter reserves. Trees with dense growth and crossed limbs that shade one another are inefficient solar collectors. Remember, the fruit-bearing wood must receive the equivalent of at least 30 percent of the available sunlight in order to initiate flower buds. This won't happen unless you prune carefully. If you've ever seen an older, unpruned tree at harvesttime, you may have noticed that the fruit was concentrated at the tips of the branches and toward the top of the tree, the places that receive the most sunlight. Perhaps you also noticed that there didn't seem to be a great deal of fruit. Don't undermine your trees' potential—the amount of time you spend carefully pruning is balanced out at the other end of the season, when you get to spend lots of time harvesting a nice, big crop.

THE CENTRAL LEADER SHAPE

Determining the ideal size and shape for fruit trees has been a goal of scientists for years, but they still have a lot to learn about this subject. Each fruit tree species has its own style of growth, which can also vary from variety to variety. But for our purposes, we can pick a model of what we'll call the "ideal" fruit tree to understand the basics about pruning and training.

The most efficient tree shape for gathering light is a cone with a single, straight trunk and branches that come out at an angle of 60 degrees. This is called a central leader tree. A nicely shaped Christmas tree is a good example of this shape. Apple trees can be pruned to central leader. Some varieties of dwarf apples naturally grow close to this ideal form. Others, such as the popular Red Delicious, need quite a bit of persuading on the part of the grower. Since the branches of central leader trees tend to grow more upright, they need to be spread to be coaxed into the best configuration. And because of their rapid growth, they need persistant pruning. But considering the payoff in yield, pruning and training your trees correctly is worth the extra work.

THE OPEN CENTER SHAPE

It's ironic that a vast number of fruit trees can't be grown in the ideal central leader form. These trees naturally produce several large branches instead of one strong trunk. You should allow trees like these to develop three or four branches which project out from the center in different directions, forming what is called an open center or vase-shaped tree. This method is best for large, vigorous trees such as sweet cherries, since it helps keep them from growing too tall and opens up the center of the tree to light. Apricot, peach, plum, prune and nectarine trees are all good candidates for this type of pruning.

A CRASH COURSE ON THE PRINCIPLES OF PRUNING

Before we discuss the ins and outs of pruning and training, we need to define some basic terms. You already know that a modern grafted tree consists of a rootstock and a scion. The scion has several parts. The trunk is the main axis of the tree which is joined to the roots. The area where two branches join, or where a branch comes off the trunk, is called a crotch. A leader is a vigorous upright branch, so the central leader is really just an extension of the trunk from the base to the very top of the tree. Limbs that come directly off the trunk are called primary scaffold limbs, while secondary scaffolds arise from the primary scaffolds. Laterals are smaller branches that appear on both primary and secondary scaffolds. A spur is a short branch with closely spaced nodes. The buds that form on spurs can produce flowers and/or leaves. A water sprout is any vigorous, upright shoot that grows from the trunk or from the scaffold limbs. Another distinguishing characteristic of a water sprout is that the wood tends to be much greener than on other parts of the tree. And finally, at ground level, a sucker is an upright shoot that grows from the rootstock.

A very important part of pruning is using the proper equipment. You'll need a good, sharp pair of hand shears or snippers for cutting off twigs and small branches and a pruning saw or lopping shears for tackling larger limbs. It's important that your tools be sharp so they can cut cleanly. If they leave any jagged bits of wood or torn strips of bark that can peel back, those are inviting places for disease organisms to enter.

PRUNING CUTS

There are two basic types of pruning cuts you can make. When you remove an entire branch, you're making a thinning-out cut. Cuts that just shorten a branch

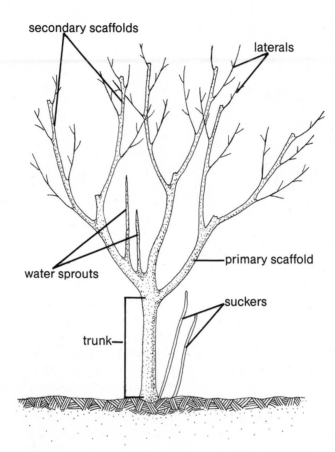

Anatomy of a Tree: In order for you to be able to prune and train your fruit trees correctly, you should become familiar with the important features that are labeled on this open center tree.

are called heading-back cuts. When you make a thinning-out cut, you don't upset the movement of growth regulators (mostly the hormone auxin) in the remaining tree. Since you're removing the entire branch, you're taking away all the buds that produce auxin. That's the reason why thinning-out cuts don't encourage a flush of vigorous vegetative growth.

You should make a thinning-out cut where the branch begins to narrow close to the trunk. Right where the branch comes out of the trunk you'll see a thickened area that slants away from the branch. This is called the bark ridge. If you were to draw a line straight down from the bark ridge, you'd see a thickened area at the base of the branch. This is the branch collar. Inside the collar is a chemical protective layer as well as special cells that help in the wound healing process. If

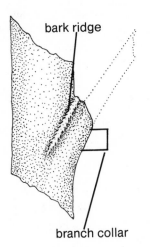

bark ridge

branch collar

Well-Placed Thinning Out Cut: To get your bearings, start your cut at the bark ridge. Instead of sawing straight down, which would remove the beneficial branch collar, saw at an angle so you come out at the point where the branch begins to narrow above the branch collar.

you cut the branch flush with the trunk or scaffold limb, you will remove the branch collar, and the wound will take longer to heal. If you go to the other extreme and leave the branch too long, the cut will be beyond the reach of the special healing cells and won't close over properly, leaving a place for disease organisms to enter.

Heading-back cuts do upset the distribution of growth regulators in a limb since they leave many of the auxin-producing buds in place. When you trim away a portion of the branch, the buds near the cut start to grow vegetatively and produce auxin. The effect of the auxin can be strong enough to cause buds that would otherwise have formed fruiting spurs to develop into vegetative lateral branches instead. When you're pruning a fruit tree, one of your goals is to encourage the tree to form fruiting wood rather than vegetative growth, especially when the tree is young. Since heading-back cuts lead to more vegetative growth, you usually want to do as much of your pruning as possible with thinning-out cuts. This will give you a smaller, more fruitful tree. But there are exceptions; as you will see later, there are times when pruning peaches and sour cherries that you need to make heading-back cuts to encourage the growth of new fruiting wood.

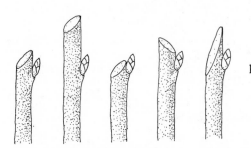

Right and Wrong Way to Cut: When you make a heading back cut, follow the example on the left. Make the cut close to but not right at the bud and angle it so it slopes away at roughly 45 degrees. If the stub is too long (second branch), disease problems may develop. A cut that's too close (third branch) will probably kill the bud. A cut angled towards the bud (fourth branch) causes water to collect at the bud which can encourage disease. When a cut is angled too sharply (far right), the wound won't heal properly.

Although heading-back cuts do stimulate growth, a pruned tree will still grow less than an unpruned one. (This is something that growers with limited land and closely spaced trees should remember.) While branches near the cuts will grow at a rapid pace, the overall growth of the tree will be slowed. Whenever you cut off branches or portions of branches, you're removing both stored food and areas that would have carried food-producing leaves. Pruning is important and it's something that you need to do to get your trees to bear well—but don't go overboard. Limit your pruning to what is necessary to shape the tree and open it to sunlight. If you trim away too much, you'll end up doing your tree more harm than good.

GETTING THE TIMING RIGHT

Each year, you should do the necessary pruning in late winter, while the tree is still dormant. Limited summer pruning is all right only under special circumstances. When you prune during the summer, you are removing photosynthesizing leaves, so you can set the tree back. If you prune in the fall, the tree doesn't have time to heal before winter and will be very susceptible to cold damage. For a tree planted in the fall, hold off on doing the initial pruning we recommend until late in the winter.

When you plant a tree in the spring, you should do the pruning right away. If you delay for a while and don't get around to it until after the tree has put out leaves, you'll be wasting your plant's energy. All the effort it put into developing those leaves will be lost with a couple of snips of your pruning shears.

PRUNING A CENTRAL LEADER TREE

The time to start shaping your tree into a central leader form is right when you plant it. If your tree is a whip (one year old) with no branches, head it back to 30 or 36 inches, or the height at which you want the first scaffold limb to come off the trunk (we give the specifics for individual fruit trees in the following chapters). This first cut encourages a vigorous central leader to grow with scaffold limbs that appear at the proper height. If you head the tree too low, the branches that develop will be much closer to the ground than you want them to be. But if you head the tree too high, scaffold limbs won't form within easy picking and pruning range.

Be sure to make the heading cut just above a strong bud. You can recognize one of these by its nice fat shape and its healthy green color. Never choose a bud that's small, blackened or shriveled. Heading back will encourage lateral branching of scaffold limbs and will also keep the top shoot in balance with the reduced root area caused by normal transplanting damage. If you remove the next three buds below the top one, you'll spur the central leader on to bigger and better growth. Wait, however, until the top bud has started to grow before you remove these lower ones. If you jump the gun and the top bud turns out to be less than vigorous, you won't have any back-up buds in the immediate vicinity to take over. Simply pinch the buds off or roll your finger firmly across them to snap them off. Not only will this remove competition for the leader, but it will also encourage the scaffold limbs to grow lower down on the tree where you need them.

SIZING UP THE SCAFFOLDS

If the tree you're starting out with has branches, you must decide whether or not to keep them. Depending on how far down you want to bend when you pick your first fruit, you'll want the lowest scaffold branch to be anywhere from 18 to 36 inches from the ground. Thin out any limbs lower than your first scaffold.

Next, step back and take a critical look at your tree to determine which branches to keep for the remaining scaffold limbs (you'll be using these same criteria throughout the tree's life to choose which of the developing branches to retain as scaffolds). You want your scaffolds to be at least 8 to 12 inches apart vertically along the trunk. This may seem like a lot of space to allow, but remember, they will increase in diameter and will send out laterals as the tree grows. Another way to select the best branches for scaffolds is to look down on the top of the tree. Ideally, the scaffold limbs should reach out from the trunk at about 90 degree angles in relation to each other. If you imagine the trunk as the center of a clock, the scaffold limbs should emerge at 12, 3, 6 and 9 o'clock. The first year, your tree should have no more than two or three scaffolds. From then on, the number of scaffolds you choose each year depends on how much growth the tree has made and how many new branches have appeared.

SPREADING LIMBS

Although it isn't always possible, you should try to select scaffold limbs that come out from the trunk at an angle close to 60 degrees. If an otherwise perfect candidate is growing at a sharper angle (closer to the vertical) and you really don't have another likely prospect for a scaffold, you can remedy this by spreading the limb to improve the angle as it grows.

There are some very good reasons why you should avoid branches with sharp angles. Excess bark will grow in the crotch, producing an area that is highly susceptible to diseases. This bottleneck also prevents an even deposit of wood where the limb joins the trunk, making for a weak connection. But if you spread a limb to a 60 degree angle, the wood that's deposited each year will be evenly distributed around the juncture, strengthening it with time instead of weakening it. Branches with narrow crotches are also slower to harden in the fall, because the sharp angles of the conducting tissue slow down the movement of nutrients. This makes them more likely to be winter-killed.

Besides improving the tree's structure and hardiness, limb spreading also makes it more efficient at gathering light. When the branches are at 60 degree angles, the sun will penetrate very close to the center of the tree, exposing the leaves along the entire length of the branches to light.

And last, but not least, when you spread tree limbs, you encourage early flowering. The hormones that trigger flower buds to grow don't flow out of branches

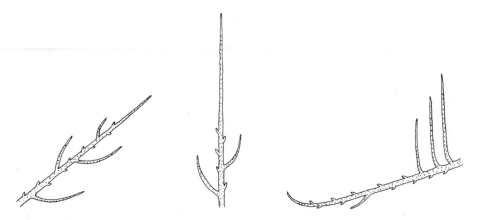

Three Kinds of Limb Growth: A branch that's properly angled at 60 degrees (left) will produce a good mix of fruit buds and lateral branches. One that's allowed to grow vertically (center) will develop very few fruit buds and lateral branches. On a branch that dips below the horizontal (right), there will be lots of fruit buds, but along with them more water sprouts. Fewer lateral branches will appear.

growing close to the horizontal as fast as they leave more vertical branches. You want to keep these hormones in the branches, not let them escape to other parts of the tree. A tree with well-spread limbs will blossom earlier than one with more upright branches. By promoting early blossoming, you're establishing a positive cycle for your tree's growth. Early fruiting will cause more of the tree's energy to go into fruit production, and it will grow more slowly. This slow vegetative growth will in turn encourage blossoming the next year when the whole cycle begins again. (Of course, if you let your young tree bear so much fruit that it becomes exhausted, both vegetative and fruiting growth will be set back.) Slowing the vegetative growth also allows the tree to remain efficient at gathering sunlight since fewer and shorter branches will be produced. This in turn means you need to do less pruning!

All of the individual benefits of spreading combine to have a significant influence on fruiting. When scientists at Auburn University in Alabama compared the yield of apple trees with scaffold limbs kept carefully spread with that of trees with limbs that were spread at first and then not touched for five years, they found a notable difference. The trees with limbs that were continually spread produced 12 to 13 percent more apples than the others. These numbers should be enough to convince you that spreading is something you should pay attention to, not just once, but throughout the lifetime of your central leader tree.

The easiest way to encourage branches to grow at a better angle is to reach into your laundry basket for a clothespin (the wooden kind with the little spring is best). Clip the mouth of the pin onto the trunk and let the top push the limb into place. Clothespins work well on new shoots that are 4 to 6 inches long. You only need to leave them on for a couple of weeks. If you forget about them, they can start

Cheap and Handy Spreader: The easiest way to nudge a young shoot into the proper angle is with a clothespin. Clip the mouth onto the trunk and let the long ends hold the shoot in place.

to girdle the branches, and then it becomes a struggle to get the clothespin off without damaging the branch.

You can make a sturdier limb spreader out of a bit of wood and some finishing nails. Push the limb to a 60 degree angle and then measure to determine the length spreader you will need. Narrow wood works the best. Remove the heads on several finishing nails and pound one of these headless nails into each end of the stick so that the pointed end projects out. Place one nail tip on the trunk and position the other on the branch so that it holds the limb at a 60 degree angle. Once the spreader has been in place for three or four years, the wood of the tree should be set and you can take the spreader out. You shouldn't have any problems removing it, but if for some reason the tree won't give up the spreader without a struggle, you'd be better off leaving it in place.

Some fruit growing books recommend using narrow pieces of wood with V's cut in each end as spreaders. We caution *against* using these because they tend to scrape away sizeable portions of bark that can make a tree an inviting target for disease. The two small holes a nail spreader leaves aren't likely to pose a problem.

Later on, if the tips of the branches tend to grow upward, you can use a rope or cord to nudge the limb back down into place. Make a loop 4 to 6 inches bigger than the branch and slip it on the limb about one-third of the way back from the tip. If you lodge it next to a lateral it shouldn't slip off. Anchor the other end to a stake in the ground.

FOLLOW-UP PRUNING — SOME DO AND SOME DON'T

Once you've completed the initial pruning and spreading of your new tree, you can sit back and watch it grow, at least for awhile. There are two philosophies about how to manage a young fruit tree during its first year. Some orchardists believe that a new tree is best left alone to grow as it will. This way, the maximum leaf area will develop, allowing more photosynthesis to take place. The tree can then produce plenty of carbohydrate reserves for the winter and for early growth the next spring. Other fruit growers feel that a young tree should be encouraged to develop its most advantageous shape during the first year while the branches are small and growth is vigorous.

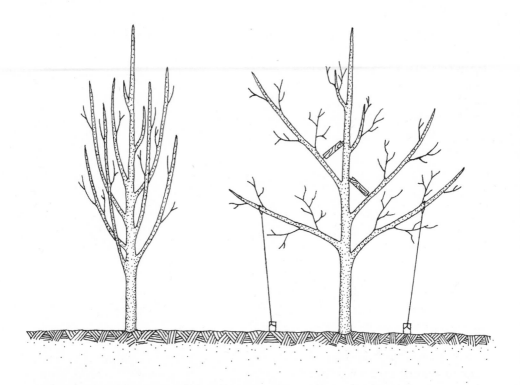

Central Leader Tree: On the left is a central leader tree that was left to fend for itself. The lack of pruning and spreading has resulted in a dense tangle of narrowly angled branches. On the right, you see a tree with carefully selected scaffolds spread to the proper angle. In the top branches, nail spreaders are in place, while the bottom branches are gently spread with anchored ropes. Thinning out cuts have opened up the tree to sunlight.

How do you determine which approach is best for you? The hands-on summer pruning approach does take time, since you're carefully scrutinizing all the new growth that appears and deciding whether to keep it or prune it away. If you can't keep up with your tree and attend to this pruning conscientiously, you'd be better off leaving it alone and waiting until it's time to do the regular dormant season pruning. But if you've got the time, summer pruning is a good thing to do. We'd encourage southern growers to consider summer pruning as a way to take advantage of their long growing season and get a head start on shaping their trees.

If you decide to direct the growth of your young tree during the first season, you'll have to check it frequently and prune it often. As the summer progresses, cut off any undesirable branches *before* they reach 2 inches in length. Removing limbs longer than 2 inches may set the tree back, so you're better off leaving them alone until it's time for winter pruning. You should cut off any branches that compete with the central leader or shade it. Also prune away any branches that grow off the trunk but won't be retained as scaffold limbs.

If you find your tree growing a group of nearly vertical branches at the top instead of a single central leader, select the topmost branch as the leader and cut the rest off at the trunk. You could retain a competing branch lower down on the trunk as a scaffold branch by spreading the limb as it grows; this would help reduce the amount of pruning you have to do. Should you notice that the central leader is drooping to either side rather than growing straight toward the sky, stake your tree if you haven't done so already (or add a taller stake if the original one is too short) and tie the leader to the stake.

PRUNING A MATURE CENTRAL LEADER TREE

While the variety of tree and type of rootstock have a major effect on how early a fruit tree bears, your pruning and training efforts also have a significant influence. You want to be careful not to overprune, but you can't just ignore your growing tree either. What we offer here are judicious guidelines to help you distinguish between too much and too little pruning.

Every year, cut off any suckers that may have grown from the roots. Leave the central leader alone after the initial cutting back, unless it has grown excessively and there hasn't been much lateral branching. In this case, cut about one-third of last season's growth from the leader. (Cut back to just above a plump bud.) When your tree gets as tall as you want it to be, each year cut the central leader back to a weak lateral branch at the height you want. Select new scaffold limbs each season, spreading them if they're growing at too sharp an angle. Also, head them back one-quarter to one-third their length to promote branching. Those new branches will become fruit-producing areas. Prune away any branches growing off the trunk which won't be scaffolds, as well as any lateral branches that are shading the scaffolds.

Keep in mind as you prune that you want the tree to be as nearly conical in shape as possible, with longer branches on the bottom and shorter ones above. This will prevent adjacent trees from overlapping one another, and it will keep the upper branches of a tree from shading the lower branches, making the ones on the bottom more productive. It's to your advantage to encourage fruit production on the lower branches, since they're easier to pick from and to train.

If top growth does begin to overwhelm the lower branches and your tree is becoming top-heavy, you should carefully consider the best way to prune. Since many small cuts tend to encourage vegetative growth more than a few large ones, your best approach is to remove one or two entire branches rather than take off many small sideshoots. When you do head back the upper branches, cut back to a weak lateral that's growing close to the horizontal or slightly upward. We usually tell you to cut back to vigorous parts, but in this case you want to contain the growth. Since the lower branches are not as vigorous as the upper ones, you need to encourage their growth. You can do this by heading back the one-year-old wood on these branches to a fat, healthy bud.

Keep your eyes out for upright, vigorous water sprouts and remove any you find. You should allow some of this growth to remain only if its removal would leave

Rejuvenating an Older Tree: Sometimes an older central leader tree will lose its shape and start to look a bit top-heavy. To restore its form, head back top branches to weak laterals or thin out one or two upper scaffold limbs. To invigorate the lower branches, head them back to healthy buds.

an underlying scaffold branch with a bare, shootless section of 18 inches or more. In that case, cut the water sprout back to four or six buds. Better yet, if it has a lateral branch growing out near its base, as water sprouts on apples often do, cut back to that lateral.

Some orchardists will summer prune water sprouts on a very vigorous tree instead of waiting until late winter. There are problems with this method, however. Summer pruning of this type should be done in August; if you prune the water sprouts any earlier, you may just encourage more vegetative growth, which is exactly what you want to avoid. But pruning in August, after most of the tree's growing is over, risks depleting its energy reserves by removing the photosynthesizing leaves and the carbohydrates stored in the wood. Late cuts can also leave the tree open to cold injury or infection by disease organisms.

Most fruit growers play it safe by waiting to do this pruning until late winter. But if you live in a mild climate and have a very vigorous tree that needs to be controlled, you can try summer pruning as long as you time it right.

If your fruit trees are overlapping or shading one another, be careful how you solve the problem. You might assume that the reasonable solution is to simply cut back the tips of the branches so that each tree has its own free, unshaded space—

but that would be the worst thing you could do! When you prune like that, with so many separate cuts on each tree, you're only encouraging a new surge of vegetative growth. That sort of growth will hamper the tree's fruiting and will also recreate the crowding problem in short order. In the long run, you'll be worse off than when you started!

The best solution to an overcrowded grouping of trees is to examine each one, pick out the scaffold limbs that are the worst offenders, and simply remove entire branches. This kind of pruning will stimulate far less vegetative growth than would extensive, smaller pruning cuts.

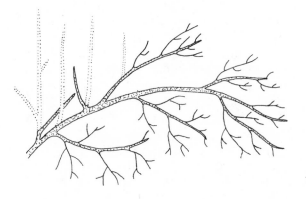

Removing Water Sprouts: As a rule, when you see a water sprout, cut it out. There is an exception. If an underlying branch will be left with a long bare stretch of wood, leave some water sprouts behind. At the left end of the limb, a water sprout was cut back to a lateral branch at its base. Another option is to leave an entire water sprout that is growing somewhat laterally, as shown to the right of the first water sprout.

SOME FINAL WORDS OF ADVICE

In all the pruning and training you do, remember that your basic goal is to produce a tree with widely spread and separate limbs that will be exposed to maximum sunlight. When you accomplish this your tree will reward you with the biggest possible crop of plump, well-colored fruit. If you pay attention to good pruning and training practices when your trees are young, you will have fewer problems later on.

By four years of age, most trees will have established their basic form; after this time, it will be difficult to change their structure. For this reason, we do not recommend trying to alter the basic shape of an older fruit tree; if it was trained in the open center form, don't attempt to turn it into a central leader tree. It would take the tree three to four years to recover from the massive pruning necessary to change its shape, and you still might not succeed. You're better off pruning out oversize branches as we suggest above for crowded trees or, if the tree is very old and in poor condition, just gritting your teeth, cutting it down, and planting a new semidwarf tree or two that you can prune and train into the form you want.

PRUNING AN OPEN CENTER TREE

With an open center tree, you want to establish several primary scaffolds instead of a single central leader. The secondary scaffolds that come off these primary ones are the branches that will bear the fruit. Actually, each of the primary scaffolds is like the central leader—it supports the fruit-bearing branches.

If you're starting out with an unbranched whip, head it back after you plant it. Open center trees are generally headed back lower than central leader trees to help keep their height down. (We give you the specific heights in the individual fruit tree chapters.) As with the central leader tree, heading back will encourage strong, vigorous scaffold limbs to grow.

If your young tree has branches near the top, look them over to see whether any would make suitable scaffolds. Eventually, you'll want to end up with three to four good, strong scaffolds, but your tree probably won't have this number so early in its growth. As the tree matures you can select more scaffolds to complete the set. What you want are branches separated from each other along the trunk by 6 to 12 inches. This ensures that as they increase in diameter, they won't all end up coming off the trunk at the same place. (Your tree will be stronger and less likely to split if the scaffolds emerge from the trunk at different places.) Ideally, you want scaffolds growing at a 45 degree angle to the trunk. If the angles of some

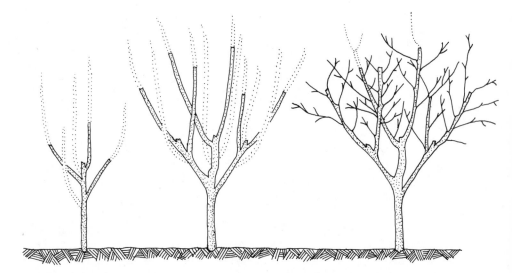

Open Center Pruning: Your first step is to choose three to four well-spaced scaffold limbs, then head them back as shown at left. At the end of the second dormant season, select secondary scaffolds, as shown in the middle view, and head them back. To maintain a mature tree at a convenient height for harvesting and tending, cut the secondaries back each year, as shown on the right.

branches are too narrow but they are otherwise good scaffolds, you can always spread them (follow the advice given in the preceding section, Pruning a Mature Central Leader Tree). Limbs growing at an angle narrower than 45 degrees won't be very strong, and those with a greater angle won't grow upward enough to give the tree its best shape.

You may be puzzled as to why we're advising a 45 degree angle here, after we recommended limbs angled at 60 degrees in the preceding section on central leader trees. The truth of the matter is that because of their vase-shaped form, open center trees can't support primary scaffolds at wide, 60 degree angles. Because 45 degrees is the best angle for these trees, they are weaker structurally than central leader trees. The way the limbs are angled, there's more of a chance that a primary scaffold will break off under a heavy fruit load or during a gusty storm. That's part of the reason why open center trees have three or four main branches—if one is lost, your tree can still bear fruit. Next time you drive by an orchard, look for trees pruned to open centers. Chances are many of them will have lost a limb somewhere along the way, but they're still able to produce on the limbs that remain.

If your young tree has only one or two suitable scaffolds at first, you need to keep these in check so they don't have a sizeable head start on the third and possibly fourth scaffolds you'll be picking out later. Head the first ones back to the lowest, outward-facing bud. If you fail to do this, these early scaffolds will always be bigger and heavier than the ones that follow and your tree will be thrown out of balance.

Should you be in the enviable position of having *all* the scaffolds there right from the start, head them all back one-third to one-half their length, depending on the type of tree (see the individual fruit tree chapters that follow for the specifics). You want each stub to end up with one or two buds. Thin out any remaining branches that you don't want to use as scaffolds. Once you've done this preliminary pruning you can either let your tree grow freely for that first season or continue to visit it, pruning shears in hand. (For the pros and cons of summer pruning, see the section Follow-up Pruning—Some Do and Some Don't.) If you do summer prune, be sure to remove any branches that start to form lower than you want your first scaffold to be before they are longer than 2 inches.

At the end of the first winter, you should select the rest of the primary scaffolds. Remove all the other branches from the trunk with thinning-out cuts and snip off any root suckers you see. With some trees, such as apricots and cherries, you'll have to do a little more pruning. These trees need to have their primary scaffolds headed back to where you want the lowest secondary scaffold branch to grow. This heading back will encourage lateral branching. When you cut, leave primary scaffolds on standard trees such as sweet cherries 24 to 36 inches long and prune to an outside secondary. With dwarfs, the primaries will be shorter—generally about 18 inches in length. Remove any crossing secondary branches or limbs that will shade those beneath them.

During the second summer, if any primary scaffolds begin to bend downward because they are carrying too much weight, you may need to remove some of the excess wood so that the primaries maintain an angle as close to 45 degrees from the trunk as possible. Other than this emergency pruning, you really shouldn't do anything else to your trees at this point.

When it's time to do the second dormant season pruning, your job is to select secondary scaffolds. On each primary scaffold, choose two to three lateral branches that appear anywhere from 24 to 36 inches from the crotch. If your tree is developing more of a closed shape than you'd like because the scaffolds are growing quite vertically, choose outward-growing secondaries to open the tree up. On the other hand, if the scaffolds are more horizontal, giving the tree too open a shape, pick secondaries that are growing more upright. Once you've chosen the secondary scaffolds, head them back to encourage lateral branching. The secondary scaffolds should be about the same length as the primaries.

Don't assume that you have to prune away all the small laterals that are springing out from the primary scaffolds. You actually want to keep some of those that are growing close to the horizontal because they, too, can produce fruit. In fact, they'll produce fruit at an earlier age than the secondary scaffolds. But as those secondary scaffolds grow older and produce more leaves, they'll block the laterals, which will eventually shut down production. This doesn't happen over-night—it can take years for the secondary scaffolds to overshadow the laterals. At that point you can prune all the laterals away.

FOLLOW-UP PRUNING

Basically, by the end of the second major pruning session you'll have established the framework for your open center tree. In the years that follow, the dormant season pruning you do is meant to keep the tree in shape—meaning nice and open so that plenty of light can penetrate to the fruiting wood. Remove any limbs that are shading others. If your tree is growing too vertically, prune to outside limbs and buds to open it up. If it's a real spreader, prune to more upright limbs and inside buds to close up the shape a bit. Once the tree reaches the height you find most convenient for harvesting and pruning, cut the scaffolds back to an outward-growing lateral at that height each year to keep your tree a manageable size.

PROBLEMS WITH FRUIT TREES

Chances are you will have to do battle with some pests and diseases over the course of your tree's lifetime. In the individual tree fruit chapters that follow, we discuss the most common problems that are likely to occur and give you symptoms to watch for, ways to solve problems, and most importantly, methods to prevent the problems in the first place. If you're a vigilant home fruit grower and check your trees often, you can hope to catch problems before they get a chance to become really serious.

The one disease that all the fruit trees discussed in this book can succumb to is verticillium wilt. (This disease is also discussed in chapter 1 under Problems with

Fruiting Plants.) When a tree becomes infected, its leaves wilt and turn yellow. As the disease progresses the leaves fall off prematurely, in the summer, before the tree has a chance to produce enough food reserves for the winter. Sometimes an infected tree will die in the first summer, but some trees can survive a couple of years before succumbing.

If you spot any of the early symptoms on your tree, don't waste any time. Immediately prune off all wilted branches and those from which the leaves have fallen. Then, fertilize the tree to make it grow vigorously. If these measures fail and your tree dies, don't replant another in the same place. Remove the top of the tree, dig up all the roots as well, and dispose of them as far away from your yard as you can. Never put any parts of an infected tree in the compost pile.

Not all fruit tree pests are insects. Deer can be a serious problem too, since they're basically browsing animals that prefer to feed on low-growing, tender shoots of trees and bushes. They also like to feast on young bark, and if they strip away a significant portion they'll upset the tree's nitrogen reserves (remember, trees store nitrogen in their bark). Be sure to give a tree that's been nibbled on a shot of compost or high-nitrogen fertilizer early in the spring.

If you surround your tree with a 4-foot wire fence, 1½ to 2 feet outside the drip-line, you should be able to keep deer at bay and not have to worry about gnawed-off bark. This fence can also protect your tree from male dogs, whose urine can damage the trunk; nitrogen from the urine can also burn the roots.

If you find a fence unattractive or impractical, you can try one of these various other remedies to discourage deer. Some people claim that bags full of human hair, blood meal or lion or tiger manure (from a zoo or circus) tied among the branches will persuade the deer to keep their distance. Keep in mind, however, that these need to be renewed weekly to maintain their effectiveness. Some growers have found a spray made of 18 eggs blended with 5 gallons of water to be a very successful repellent. If you're turned off by the thought of having your yard smell like one big rotten egg—never fear. Only the deer can smell the odor of the decomposing eggs. While it's enough to keep them away, you won't be able to detect anything.

CHAPTER 7

APPLES
TEMPTATION FOR ALL
FRUIT LOVERS

Vital Statistics

FAMILY
Rosaceae

SPECIES
Malus pumila

POLLINATION NEEDS
Some varieties require cross-pollination
while all the rest will set larger crops
with cross-pollination.

HOW LONG TO BEARING
Dwarf   3–4 years
Standard   4–8 years

CLIMATE RANGE
Zones 5–7 with some varieties
in Zones 4 and 8

Many of us who grew up in small towns have memories of big old apple trees in our neighborhoods which were favorite climbing trees. The braver children picked tangy, crisp fruit from these trees, while the less adventuresome gathered the windfalls that collected under the spreading branches. Only folks with spacious yards could have their own apple trees in those days. But thanks to the efforts of many determined and patient researchers, the home gardener today with even the smallest bit of land can grow apples. A complete range of sizes, from the 6-foot midget tree that will grow in a patio or balcony tub all the way up to the familiar full-size climbing tree, is available now. Since apples vary so much in flavor and texture, as well as in keeping ability, many home fruit growers will be tempted to grow several different varieties—something that small-size trees make possible.

Apples have been cultivated since the dawn of history. Both the Greeks and Romans grew apples, and budding and grafting of apples have been going on for 2,000 years! The cultivated apple we know today has been developed down through the ages through the cross-breeding of several different species. Generally, it is given the name *Malus pumila*. Apples evolved throughout a rather large area including Asia Minor, Central Asia, Himalayan India and Pakistan. Some of the most popular varieties today are among the oldest; Pippin apples, for example, were popular in England in the sixteenth century.

APPLE VARIETIES

Apples far exceed any other cultivated fruit in the number of varieties available—there are over 100 types of Delicious apple alone! Since it's impossible for us to discuss all apple varieties, we'll try to give you some guidelines to help you pick out those you want to grow. Refer to the Guide to Apple Varieties at the end of the chapter and to the section on apple rootstocks for useful information on specific trees.

Assortment of Apples: Feast your eyes on the wonderful assortment of apple shapes and sizes. Here's just a glimpse at a few of the varieties available. From left to right, the yellow, oblong Ortley is favored for fresh eating; the green and red Northern Spy is a fine cooking and eating apple; the yellow Newton Pippin is a good storage variety; and Arkansas Black is a striking maroon to black-colored apple that stores well.

CONSIDER YOUR CLIMATE

Before you begin to browse through catalogs and fall in love with a particular variety, avoid possible disappointment by considering your climate; where you live will have a distinct limiting effect on your choice of apples.

Unfortunately, not all nursery catalogs give the kind of information you need to make an intelligent decision. For example, the crisp, dark green variety widely grown in New Zealand, Granny Smith, has become available here in the last few years. Nurseries sing its praises and make it sound like the apple for everyone. But Granny Smith won't do well where it's either very hot or very cold. This variety blooms early and requires a very long growing season before the apples are ready for picking. Just to give you some idea, trees are finally ready for harvest anywhere from mid-October into November. For many growers, these apples are just too slow! Clearly, most of us will still have to buy these tart fruits in the supermarket if we want to eat them. The variety Golden Delicious, on the other hand, is very widely adapted and can be succesfully grown in all but the hottest and coldest parts of the country.

Apples are generally suited to a temperate climate and require a winter chill of 900 to 1,000 or more hours between 32° and 45°F. For a long time this has meant that people in mild-winter climates haven't had too much success with apples. But now there are tasty new low-chill varieties such as Dorsett Golden and Anna available, which can be grown as far south as the Gulf Coast. Make sure when you choose your apples that you don't pick a variety that requires more chilling hours than your climate provides. If you do, the tree either won't come out of dormancy, or its growth will be late in starting and generally weak.

On the other hard, if you have a short, cool growing season like we do here in Montana, be sure to pick hardy varieties that don't take too long to ripen. When Dorothy chose her Roxbury Russet tree, she wasn't aware that it was a late variety since the catalog she ordered from neglected to give that vital information. As a result, she'll rarely get the chance to savor that firm, crisp apple from her garden. In short-season areas (90 to 120 days), it's best to choose varieties labelled as midseason; these will bloom late enough to avoid any lingering spring frosts but will ripen early enough to harvest before hard fall frosts hit.

PROVIDE FOR GOOD POLLINATION

Even after you've narrowed down your choices, chances are you will still have a good selection of varieties to choose from, unless you live in the desert or the Deep South. Now you should take into account which varieties will pollinate one another effectively. It won't matter how well adapted your trees are to the climate; if you don't choose varieties which can pollinate one another, you won't get any apples.

Those of you who only have enough room for a single tree shouldn't abandon all hope of ever getting apples. Providing for good pollination doesn't always mean that you must plant two trees. Neighboring apple or crab apple trees 80 feet away or less could pollinate yours if they are compatible. Lacking nearby pollinators and the room for more than one tree, you can plant one of the varieties that is self-pollinating, such as Golden Delicious, Jonathan or Yellow Transparent. (You should know that even these varieties will produce a more abundant crop if they're cross-pollinated.)

When you're looking for suitable pollinators, be aware that there are certain apples that can't pollinate either themselves or other varieties. The popular varieties Gravenstein, Mutsu and Winesap fall into this category, as well as the less widely grown Ein Shemer and Sir Prize. The cells of these trees have an extra set of chromosomes which makes their pollen sterile; so they must be planted with another variety. If, for the second variety, you choose an apple such as McIntosh which also requires a pollinator, then you will need to plant at least three apple trees. To give you an example, a Gravenstein would require another variety such as McIntosh to pollinate it, and then a third, perhaps a Red Delicious, would be needed to pollinate the McIntosh (the Red Delicious would also help out with the Gravenstein).

There's another complicating factor in this business of cross-pollination. Some varieties are incompatible with one another and simply won't cross-pollinate. If you planted Granny Smith and Tydeman's Red, for instance, each would do some self-pollinating, but neither could pollinate the other. For a good crop, you'd have to pair the Granny Smith with a variety like Golden Delicious or Idared.

One other factor to keep in mind is that different versions of the same basic variety won't cross-pollinate. If you planted a row of different forms of Red Delicious along your fence, you would get no fruit unless you had a neighbor growing apple varieties other than Red Delicious. The same thing goes for different forms of McIntosh.

If you choose only two apple varieties, don't pick a very early one and a very late one—there may not be enough overlap in their blossoming period for them to pollinate one another adequately. Golden Delicious is a good variety to plant as a pollinator, for it produces three sets of blossoms and blooms for a long time.

Once you have settled on the varieties to plant, you need to decide what size trees you want to grow. If you're in the market for a space-saving tree, you'll have no trouble finding one. In addition to the apple trees of various sizes grafted onto dwarfing rootstocks, a few are available in compact spur-type varieties. Because they have more fruiting spurs, they can produce twice as many apples during the first ten years as standard trees.

DIFFERENT APPLES FOR DIFFERENT USES

Some people prefer to eat apples fresh and especially enjoy their crisp, crunchy texture. Others are fans of apple pie or applesauce and want an abundant crop of apples which can be cooked up in one way or another. Luckily, many apple varieties, such as Northern Spy, produce all-purpose fruits which can be enjoyed either way. But there are some varieties that are better fresh and others that are at their peak when baked in a pie.

In general, very sweet varieties, such as Red Delicious, are best for fresh eating and taste flat when cooked. Early varieties, such as Lodi, tend to be soft and a bit tart. These aren't the best qualities for fresh eating but they do make a wonderful applesauce. Soft apples tend to disintegrate too much when baked or made into a pie. The best varieties for baking and for pies are those with a firm texture and with some tartness. An exception to this is Rome Beauty, which can be mealy when eaten fresh but shines when baked or made into a pie.

In a taste-test of 12 apple varieties, *Food and Wine* magazine came to the conclusion that the best variety they tested for fresh eating was York Imperial, with Winesap and Idared tying for a close second. They also felt Granny Smith was good fresh when not underripe. For baking, Jonathan came out on top, with Golden Delicious, Rome Beauty and Granny Smith also in the running. In the sauce category, Cortland was the winner, but Empire was also considered to produce a moist, fine-textured product. And finally, the variety the testers preferred for the American classic, apple pie, was Jonathan. The runners-up were Rome Beauty, Granny Smith and Idared.

In the *Food and Wine* testing, Red Delicious didn't score many points, even fresh. This is probably because the apples were store bought rather than homegrown. When they're allowed to ripen properly on the tree (which rarely happens with commercially grown fruit), Red Delicious have a special rich flavor and a fine, crunchy texture that's unique among apples.

APPLE ROOTSTOCKS

Apple trees are grafted onto different rootstocks, each of which has its own special characteristics. Since the rootstock can make the difference between success and failure in growing apples, you should learn something about the characteristics of the kinds most commonly used.

SEEDLING ROOTSTOCKS

Standard apple trees are simply grown on seedling rootstocks. These rootstocks come mainly from the Pacific Northwest apple growing centers and usually have Red Delicious as one of the parents. Seedling rootstocks have some tolerance to fireblight, collar rot and woolly apple aphids, and they are somewhat winter hardy. But their big disadvantage is that they form very large trees that don't fruit until they are six to ten years old. Such trees are really inappropriate for home gardeners now that dwarfs are available, and even commercial growers are moving away from these hard-to-pick, full-size monsters. Since some varieties, when grafted onto seedling rootstock, reach their best yields only after 20 to 30 years, you can see why commercial growers are questioning the economic soundness of growing such trees and are switching over to the smaller, earlier bearing varieties.

CLONAL ROOTSTOCKS

Most apple rootstocks you find today are clonal ones, bred especially for certain characteristics such as dwarfing and early bearing. While there are currently rootstock breeding programs going on all over the world—most notably in England, Russia,

Poland and the United States—most modern clonal rootstocks were developed at the East Malling Research Station in England, starting in 1912. This explains why many rootstocks have names such as Malling 9. Unfortunately, there's no standard way to designate the names of these particular British rootstocks, so different forms are used by different catalogs. When you see names like EM9, EM.9, EMIX, M9, M.9 and so forth, they all refer to the *same* rootstock. The one part that stays consistent is the number itself, although it may be written as a Roman numeral. A few numbers are preceeded by MM—these are the Malling-Merton rootstocks (a joint venture between East Malling and the John Innes Horticultural Institute of Merton, England). In this book we've used the simplest way of designating rootstock numbers—MM106 and M9, for example.

Tips on Choosing Dwarf Apple Trees

When you're choosing dwarf apple trees, the very first thing you should do is check your garden soil. It's essential that you know how deep the topsoil is and whether or not it's well drained. Only then can you choose trees with rootstocks adapted to your particular conditions. Some rootstocks can tolerate brief periods of soggy soil, while others won't grow properly in shallow soil with a hardpan layer. If your soil is really deep and rich, even the smaller dwarf trees will grow more vigorously than they otherwise would, so you may need to allow for this and space them further apart than is generally recommended.

Although they have made life a lot simpler for fruit growers, the clonal rootstocks do have their problems. If you know in advance what they are, however, you can avoid disaster by not buying trees grafted onto rootstocks that could fail in your area. Besides disease susceptibilities, clonal rootstocks often develop a shallower root system than seedling types.

M9 is the most dwarfing of all the rootstocks on the market. When you buy a dwarf tree that can be grown in a tub, it's likely to be on M9 rootstock. Because their roots are brittle and they lack extensive feeder roots, M9 trees must be watered regularly or mulched heavily if planted in soils without much water-holding capacity.

Most M9 rootstocks are infected with four or more latent viruses. The rootstock has the ability to carry the viruses without being harmed by them. However, the viruses may be transmitted to the scion grafted onto the rootstock where they can affect the top part of the tree. For example, yields of Golden Delicious grafted onto M9 can be very low if it's infected by rubbery wood virus carried by the rootstock. You may wonder, then, why an effort isn't made to use virus-free rootstocks. When virus-free selections of M9 were tested, they produced trees that were more vigorous than those on infected roots, and so were less dwarf. Ironically, virus-free M9 would actually be a liability for most varieties when a very dwarf tree is desired.

An interesting array of pests is attracted to M9 trees. These trees tend to form fledgling roots above the soil surface which act as points of entry for insect pests such as trunk borers. Their bark is especially thick, and mice appear to delight in munching on it. (Heed these words of warning and be sure to use tree guards on any M9 trees you plant.) Because the trees are so small and the fruit so low, pheasants, rabbits or squirrels may discover the apples and eat them, while deer will enjoy browsing on the branches. Since they grow slowly, M9 trees will suffer more than others from deer damage.

Trees grown on M9 bear fruit very early, often within the first two or three years after planting. If you don't stake an M9 tree, a heavy fruit load may pull it over! Make sure the stake is tall enough so that the central leader can be tied to it, for the leader is likely to droop. You should also be aware that M9 appears to be highly susceptible to frost injury. In general, this rootstock is best for trees that will grow in tubs, where the trees can be easily staked and watched over. Don't select this rootstock if winter temperatures in your area drop below −20°F.

The M26 rootstock produces trees about 8 to 10 feet tall. Unlike M9, it is virus-free. M26 seems to be more compatible with Red Delicious than M9. Trees growing on M26 roots must be planted in well-drained soil, for this rootstock cannot tolerate wet feet. Although M26 produces better anchorage than M9, it also encourages early fruiting so you should be prepared to stake the tree.

A tree on M7 rootstock usually grows 8 to 10 feet tall, the same range as M26, although M7 trees tend to be somewhat taller. Some trees on M7 will even reach 12 feet if they're in rich soil. Like M9, M7 rootstock may carry latent viruses. It produces semidwarf trees consistently when combined with Golden Delicious and Idared scions, but with other varieties, tree size varies with soil conditions. To offset the poor anchorage this rootstock offers, the scion is often grafted high up on the rootstock. This means the tree can be planted up to 6 inches deeper than it grew in the nursery, which encourages more roots to grow, improving anchorage. If your soil is very deep and well drained, you can grow M7 trees without staking.

M7 can take wetter soil than M26, but don't plant it where there is an underlying hardpan. This rootstock often has a taproot; if it hits a hard layer of soil, the growth of the rest of the tree will be inhibited. You should also be aware that M7 rootstocks tend to sucker profusely.

Another Malling rootstock, M2, produces a tree about 65 percent of normal height, but the tree will only bear about one year earlier than standard trees. M2 has a good root system, however, which grows well in a variety of soils, so you don't need to stake the trees.

An MM111 rootstock will give you a tree about the same size as one grown on M2. And like M2, it has a good root system and grows well in most soils. Trees on MM111, however, sometimes grow taller than expected, which can come as a nasty surprise if you're cramped for growing space. This rootstock is good for northern areas, for it's quite winter hardy. In addition, MM111 is very drought resistant and can tolerate wet soils, as well. When it comes to the harvest, there are some drawbacks; MM111 doesn't favor early fruiting, and trees on this rootstock give a smaller harvest than trees on other rootstocks such as MM106.

The Importance of Deep Loamy Soil: If you plant a fruit tree where there's a layer of hardpan, you're asking for trouble. The roots will be confined and when that happens, the topgrowth will be stunted. Hardpan layers are common in soils with a lot of clay or where mineral salts have built up.

Researchers bred MM106 especially for resistance to the woolly apple aphid. This rootstock produces trees that vary greatly in size, from 65 percent of full size to almost full size, depending on the variety and on growing conditions. Trees grafted onto MM106 come into bearing early and they bear heavily. In hardpan soils, the roots will tend to grow horizontally, and the tree should be staked. MM106 is not a good rootstock for the North, for trees growing on it tend to break dormancy early in the spring and harden off late in the fall, making them very susceptible to frosts at both ends of the growing season.

THE ROOTSTOCK-SCION RELATION

Apple rootstocks can have some interesting effects on the fruiting part of the tree. For example, the varieties Empire, Idared, Jonathan, McIntosh and Monroe grown on seedling and certain other rootstocks will produce blossoms only on spurs at least a year old. But if these same varieties are grown on M9 or M26 rootstocks, lateral fruit buds will appear on last year's growth as well as on spurs. These buds open three to seven days later than the normal spur buds. In an area where there are late frosts, this can be a real advantage, so northern growers should look for those varieties specifically on M9 or M26 rootstocks. But in warmer climates, the late-opening buds can provide an entry site for fireblight bacteria, which thrive in warm weather. More southerly growers would be better off avoiding those trees.

While each type of rootstock tends to produce trees of a certain size, the actual height of your tree can sometimes depend as much on the scion variety as it does on the rootstock. Two different scions grown on the same type of rootstock can turn out to be different sizes. For example, a Golden Delicious tree will be shorter on an M7 rootstock than a Red Delicious growing on the same rootstock.

THE IMPORTANCE OF APPLE SPURS

Before we can start talking about apple flowers in great detail, you should know where these flowers, and ultimately the fruit, appear on the tree. Apples and pears have something in common—they both produce their fruit on short compressed branches called spurs. It's interesting to learn how these special fruit-bearing branches develop and how they affect the way fruit appears on the tree.

In the tree's second year of growth, some of the buds on a branch may develop into spurs. These spurs go on to form buds later that season at their tips. The buds are

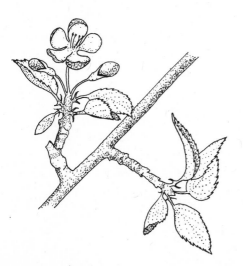

Apple Spurs: These short, compressed branches are the centers of fruit and leaf development. Each spur has the potential to produce both flowers and leaves. A spur that only produces leaves will be straight like the one on the right, while a spur that has blossomed will take on a zigzag appearance like the one on the left.

mixed; that is, they will have the potential the next spring to produce a cluster of both leaves and flowers. The flowers, if they're properly pollinated, will develop into fruit. Sometimes four or five fruits will set from one flower cluster. If you've been following the timing of this fruiting process you've seen that apples mature on spurs on main branches that are in at least their third year.

Every spur has the potential to produce both flowers and leaves, but this won't necessarily happen every season. Spurs that are bearing fruit don't form flower buds for the next year, since they're devoting all their energy reserves to the ripening fruit. For this reason, any given spur can only produce apples every other year, but it will still produce leaves every year. A spur that has fruited will spend the next season developing a tiny branch from an axillary bud. At the end of that little branch the bud that will become apple blossoms and leaves will form. This continual process of putting on tiny branches explains why an older spur that has carried fruit for several seasons bears a zigzag appearance. Sometimes outside factors like poor weather, nutrient stress or radical pruning keep a spur from bearing fruit. When this happens, the spur grows straight, not zigzagged. No matter whether a spur is fruiting or not, the nodes grow very close together, keeping these specialized branches from getting very long.

Apples are tremendous producers of flowers, which is evident to anyone who has even seen an apple tree in full, glorious bloom. But before these flower buds can develop, the level of carbohydrates within a spur must be quite high. Sunlight is the key to carbohydrate production—the leaf clusters on a spur must bask in full sunlight in order to transform the spur's potential to develop fruit buds into reality.

This is one reason why proper pruning of apple trees is so important. If the foliage is too thick or the branches too close together, not enough sunlight can reach the vegetative spurs, and they will never become fruiting spurs. The other extreme, too few leaves, can also hamper fruiting. This situation can arise when trees are overpruned or when they are defoliated by diseases or insects. Trees that are missing a sizeable number of leaves won't produce as many apples the next year as they normally would have because the level of carbohydrates in the spurs isn't high enough to promote flowering. Keep this cause-effect relation in mind and don't hinder your tree's productivity—be judicious in your pruning and do everything you can to keep pests and diseases from stripping the leaves off your trees.

APPLE BLOSSOM TIME

If you've ever paused for a moment and examined an apple tree in bloom, you probably noticed that each flower was made of 5 whitish to pink petals. What you probably didn't notice was that the same flower had 5 sepals, 20 stamens and a pistil divided into 5 styles, each with 2 ovules. (You can see these inner parts in the apple flower illustration in chapter 1.) A simple calculation reveals that the most seeds an apple can have is ten. Like other fruits, the apple seeds produce hormones which promote fruit growth, so an underpollinated apple blossom will produce a small or misshapen fruit.

BEES AND APPLE BLOSSOMS

Because apple trees put out such a profusion of flowers, and because only 10 percent of them need to be pollinated for a good crop, you might assume that there's no reason for apples to have pollination problems. But certain varieties may develop problems, and this has to do with the way honeybees collect nectar. For many years, scientists puzzled over the low set rate of Delicious apples, which often produce only half as much fruit per tree as they theoretically could. Then, in 1979, a scientist at Cornell University carefully examined Delicious flowers. He found a defect in their structure which explained the frustratingly low set rate of these trees.

Before you can understand the significance of this discovery, you need a quick lesson on how bees collect nectar from flowers. Basically, bees can get at the nectar from three different positions on the flower. The nectar is located in a circle between the bases of the styles and stamens, called a nectary. A bee can stand on the stamens

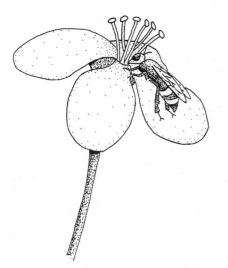

Side-Working Bee: Certain apple blossoms allow a bee to stand on the petals and reach in between the stamens to get at the nectar. Although the bee's happy, you won't be, for there's very little chance that pollination will occur.

and put its head down into the nectary to get at the nectar; then it will almost surely get pollen on its body which it will carry on to the next flower. It can also spread the flower apart by standing with some of its legs on a petal and parting the stamens with its head and/or front legs. Again, this bee will be touching the stamens and is likely to get pollen on its body.

The third way of getting nectar, called side-working, ties in with the low set rate of Delicious apples. In side-working, the bee stands on the petals and reaches into the nectary with its tongue. Since the bee isn't likely to touch the stamens in this position, flowers are rarely pollinated. Normally, side-working is difficult because the stamens form a circle that is hard for the bee to push through. But in the Delicious flower there are gaps between the stamens which the bee can easily reach through to get at the nectar.

The Cornell scientist found that 86 percent of the bees gathering nectar from Delicious flowers were side-workers, while only 6 percent of the bees collecting from 5 other varieties were. He went on to examine the flowers of 40 other varieties and found another—Northern Spy—which had flowers like those of Delicious trees. Since Northern Spy is also a shy bearer, it is probable that bees also side-work its blossoms.

While you can't alter the structure of these flowers, there are a few steps you can take to encourage better pollination. If you have trees of either of these two varieties and find your crop disappointing, you might take up beekeeping, which will increase the number of bees in the area collecting pollen to feed their brood. The more bees you have visiting your trees, the greater chance there is that pollination will occur. Another option is to plant a companion apple tree that is a good pollenizer and that begins to bloom a bit earlier than your problem tree.

This suggestion is based on another study that showed that bees are creatures of habit when it comes to gathering nectar. If Delicious flowers are the first to bloom and bees side-work them, the insects will tend to keep side-working the flowers of later blooming trees as well. This can be a problem, especially if you're growing a popular combination of apples—Delicious and Golden Delicious. While a side-working bee can be discouraged by the normal flowers of most apples, it may find Golden Delicious easy to side-work. So, if you grow both of these varieties and the Delicious blooms first, the bees will begin side-working these flowers and just continue with this less effective method when the Golden Delicious flowers open up. This could leave you with a disappointing harvest from *both* varieties! But if you plant an early blooming tree that isn't likely to be side-worked, the bees will already be in the habit of efficient nectar collection by the time your Delicious tree flowers. Once they've settled into one method, they're not likely to lapse into side-working.

THINNING FOR A BETTER HARVEST

You may find it difficult to go out into the backyard orchard and remove perfectly healthy fruit from your trees, but with apples this thinning process is especially important. If you have too many apples on a tree, more of them will be undersize, their color will be poor, and their flavor will probably be disappointing. When the tree has to provide for too many apples, there aren't enough carbohydrates to go around, and the fruit ends up being less sweet. This affects not only its flavor but also its storage life. Leaving the tree with too many apples can also exhaust its resources, discouraging the non-fruiting spurs from developing flower buds for the next year. This is all it takes to trigger a frustrating biennial bearing pattern (see Some Facts on Fruiting in chapter 6).

Another problem with very heavy crops is that as the apples swell, they may weigh down the branches and break them. You will have to provide some sort of support for the overladen limbs and even then, if there is a strong wind, you may still lose branches.

The tree itself can compensate for overfruiting to some extent. Apples with fewer than three of the ten seeds fertilized will drop off the tree; more of these drop from heavily loaded trees than from lightly pollinated ones.

Apple seeds produce auxins that help the fruit remain on the tree. There are two periods in the apple's life when auxin production is low and the fruit can fall. The first is right after the initial swelling of the young fruit, at the beginning of seed development. This is called early drop. Later on, when the embryo within the seed is developing rapidly, auxin production slows, and the June drop occurs. If there's a prolonged period of cool, wet, cloudy weather during the early development of the fruit, the leaves won't be able to produce enough nourishment for rapid seed development, auxin production will decrease, and you'll notice a heavy fruit drop during both stages.

Even though the tree is dropping fruit on its own, you'll still have to help it along. As a guideline for thinning, realize that each apple will need from 30 to 40 leaves to provide it with enough energy to grow and mature properly. (Delicious apples are the exception — they require 40 to 50 leaves per fruit.) Another way to gauge how much to thin is to allow 6 to 8 inches of branch for each fruit. Because you don't want the apples to rub against one another, thin each cluster to one apple per spur, leaving the largest one. (This can also help prevent serious codling moth problems, since these pests tend to thrive in overcrowded conditions). When you thin, be very careful not to damage the spurs from which you're removing the apples, for without healthy spurs, you won't get a crop in the future. Twist the fruit off rather than yank it from the spur.

APPLE PRUNING POINTERS

The best growth pattern for an apple tree is the central leader. When you buy a new apple tree, it will look like an unpromising stick, probably with some twigs poking out from its sides. Don't let your fledgling tree's rather homely appearance discourage you. As it grows and puts out new and vigorous branches, you can select the best limbs to provide the initial scaffold. (For a refresher on how to do this, see Pruning a Central Leader Tree in chapter 6.) The main objective of your early pruning and training is to develop a tree that is structurally able to carry the weight of many apples and that has the openness to allow sunlight to reach as many leaves as possible. The trick is to achieve these goals with as little pruning as possible.

As the tree matures, there are certain growth patterns you should watch for when you do your dormant season pruning. Cut any drooping branches back to a lateral that's growing close to the horizontal. Also remove any small branches that grow downward. Fruit growing on these branches would be shaded and thus tiny and unevenly colored.

Some apple varieties, such as Rome Beauty and Cortland, may produce lateral branches that are finely divided at the tips. This fine wood can shade itself and other

Schedule of Apple Pruning Tasks

FIRST YEAR

At Planting

Task 1:   Head back unbranched trees to 30 or 36 inches. Head back branched trees to first vigorous bud.

Task 2:   Thin out any branches that don't belong in a permanent scaffold pattern. Leave at least 8 to 12 inches vertically between scaffold limbs. Plan on a branch in each quadrant. Thin out any branches lower than 18 to 36 inches from the ground.

Task 3:   Spread scaffold branches so they form a 60 degree angle with the trunk.

During the First Summer

Task 4:   Remove any branches that compete with the central leader before they reach 2 inches in length.

SECOND YEAR AND THEREAFTER

Task 5:   Follow the same directions given for Task 2 above.

Task 6:   Follow the same directions given for Task 3 above. If branch tips tend to turn upward, tie them down to keep them at about 60 degree angles.

Task 7:   Remove root suckers, small downward turning branches and water sprouts. (If pruning out a water sprout would leave 18 inches or more of a branch without any shoots, cut the water sprout back to four or six buds or back to a lateral branch growing out near its base.)

Task 8:   If you notice any drooping branches, cut them back to a lateral branch growing close to the horizontal.

Task 9:   Keep the tree growing in a basic cone shape by removing scaffold limbs that are too large at the top of the tree or by heading them back to a horizontal lateral.

branches, resulting in fruit that is undersize and pale. Should you spot this problem on your trees, thin out the laterals, leaving an even distribution around each branch.

If you have spur-type trees, inspect them closely for branches with small buds (which will grow at a slow pace) and branched spurs. To invigorate these branches, remove the most finely divided branched spurs and leave those that have only one or two branches. When you cut one off at the tip of a branch, head it back to a large, healthy bud.

Space Invaders: If you try to squeeze too many trees into a limited amount of space, even if those trees are dwarfs or semidwarfs, they may start to crowd each other as they mature. Instead of using heading back cuts, make thinning out cuts to remove limbs that are the worst offenders.

WHAT APPLE TREES NEED TO GROW

Apple trees aren't quite as finicky as other fruit trees when it comes to the nutrients present in the soil. They aren't likely to have deficiency problems when planted in most soils with a reasonably good level of fertility, but that doesn't mean you can ignore their nutrient needs altogether. Apple trees will fail to grow and bear properly if vital nutrients are missing. In this section we'll call attention to the most important nutrients for apple trees and let you know how to tell if your trees are getting too little (or even too much) of them. Refer to chapter 1 under The Nutrients Fruiting Plants Need for symptoms of other general nutrient deficiencies along with suggestions for good organic sources of the various nutrients.

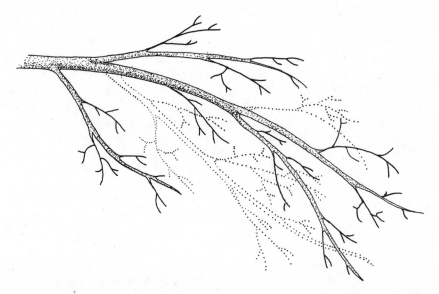

Corrective Apple Pruning: When lateral branch tips become finely divided they are likely to bear undersized, pale fruit. Remedy the situation by thinning out a few laterals in a balanced fashion. Don't take them all from the same side of the branch.

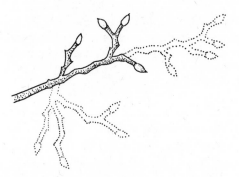

Spur Renewal: A spur-bound tree isn't going to give you much of a harvest. Inspect each branch and remove the most finely divided branched spurs. When you remove one at the tip of a branch, head back to a large, healthy bud.

Like most fruit trees, apples need nitrogen. But they can easily get too much of a good thing, so fertilizing them can be tricky. Apple trees should put on anywhere from 8 to 12 inches of new shoot growth a year, with that growth concentrated in the spring and early summer. If your tree is putting on less than 6 inches in a year, you should add more nitrogen to the soil in the early spring. If you don't, the tree will produce a measly number of flowers and set fewer, smaller apples the following year.

Trees that get too much nitrogen, on the other hand, will grow more than 14 inches in a season. An overabundance of nitrogen affects the color of the fruit, making it greener and less red than it otherwise would be. The fruit will also be oversize and mealy in texture.

Trees that bear a bumper crop every other year instead of each year will need an extra boost of nitrogen during the bearing year, as well as during the slack year. By giving such trees more nitrogen, you can hope to make them more vegetative so they'll bear less heavily and perhaps revert back to the more desirable annual bearing habit.

Calcium plays an important role in how well your apples hold up in storage. Varieties such as Northern Spy, Rhode Island Greening, Delicious and other winter storage types need to have adequate calcium, for it helps stabilize their cell membranes. Apples deficient in calcium have cell membranes that disintegrate prematurely, drastically reducing the storage life of the fruit. This disorder is called bitter pit. Apples with bitter pit have small areas of brownish, corky tissue just under the skin. You can usually spot this easily, for the skin over the affected area is sunken into a discolored pit. Eventually, the apples may be covered with these unsightly spots. If you're brave enough to taste an apple after you've pared away all the discolored spots, you'll find that it has a bitter flavor. The external signs of bitter pit may not show up until August, but apples that appear to have only minor problems when you harvest them may rapidly deteriorate in storage, depending on how serious the calcium deficiency is.

Bitter pit is not necessarily due to an overall lack of calcium; unfortunately, the problem is much more complex than that. Bitter pit is actually caused by an imbalance between nitrogen and calcium. Although nitrogen is very important for your apple tree, it can have some negative effects on how the tree takes up calcium. A tree that gets too much nitrogen may suffer from bitter pit, while another tree with less nitrogen and the same amount of calcium may be problem-free. This occurs because both calcium and nitrogen are taken up along with water into the xylem and compete with one another for the limited space available in the transport system. If there's too much nitrogen, it will overwhelm the calcium, inhibiting the tree from taking up enough of the latter nutrient.

Nitrogen given in the form of manure is available as nitrate; this ties up calcium by causing the tree to form crystals of calcium oxalate in the shoots. When many of these crystals form, they limit the tree's calcium supply so there isn't enough available for the fruit, and bitter pit develops. Ammonia, also released into the soil by manure, interferes with calcium as well. No matter in what form you supply nitrogen to your tree, too much of it can cause problems with calcium uptake.

If you have trouble with bitter pit in your apples, hold back on nitrogen, keep your soil pH at 6.0 to 6.5, and discourage your trees from the biennial bearing habit (see Some Facts on Fruiting in chapter 6). Since the fruit accumulates calcium during its early stage of development, make certain the soil is moist around the roots of your apple trees from the time flowers open until the apples appear. This will enable them to absorb calcium more readily.

A tree that's not getting enough potassium will have leaves that appear scorched. The older leaves will exhibit symptoms first, then the younger ones will follow suit. The affected areas first turn pale green, then they die. The leaf margin may be torn in places, and the leaf edges may be curled. Leaf curl generally shows up in late summer about the time the fruit begins to mature. If the deficiency is severe, new shoots may be thin and very susceptible to winter injury.

The sad fact about apple trees with a magnesium deficiency is that they drop their fruit before it's ripe. It's unfortunate that one of the most popular varieties, McIntosh, is especially susceptible to this deficiency. Acid soil can easily result in a magnesium deficiency, for apples have trouble absorbing this mineral when the soil's pH is on the low side. Researchers for the Extension Service in New York State have investigated this relation between acid soil and deficiency. After testing a variety of soils, they found that all trees grown in soil with a pH below 5.5 showed signs of magnesium deficiency; half the orchards with a soil pH between 5.5 and 6.0 had deficiency symptoms; and trees grown in soils with a pH above 6.0 showed no sign of a problem. Since 6.0 seems to be the magic number, monitor your soil's pH so that it never drops below this level and you should head off any possible deficiency.

A manganese deficiency is rare, but it can occur, and when it does it manifests itself in a unique way in Delicious apple trees. The wood becomes very rough, with small raised lumps overlying dead tissue. Because of these tiny bumps, this disorder is sometimes referred to as measles! A manganese deficiency generally leads to small terminal shoots, stunted trees and a sparse harvest.

Apples from trees with a boron deficiency develop corky material near their cores. In severe cases, the corklike tissue extends all the way through the fruit.

HOW MUCH COLD CAN APPLES TAKE?

Just because apples are the most cold-adapted of our tree fruits, that doesn't mean they never suffer from cold injury. In fact, the older an apple tree is, the more susceptible it is to cold damage. In north central Washington, researchers found that Red Delicious trees over 15 years old were 10°F less cold resistant than younger trees. This means that a cold snap could kill all the older trees and leave the younger ones untouched. The researchers also discovered that Golden Delicious trees exposed to wind and cold had a 45 percent better chance of survival if they were younger than 15 years.

The reasons why hardiness goes down as age goes up could actually be a combination of factors: an accumulation of previous cold injury; the depletion of carbohydrate reserves by the heavy crops carried by older trees; insufficient leaf area relative to the volume of wood (meaning not enough carbohydrates could be produced to enhance the tree's hardiness); and dormant disease organisms present in older trees which make it more difficult for them to recover from cold exposure. Considering all the various factors that can come into play, if you have an older apple tree that succumbs to a bad winter, you really shouldn't be too surprised.

Late frosts have varying effects on apples. If freezing temperatures occur just as the young fruits are developing, the immature seeds on one side of an apple may be killed, while those on the other side may escape unharmed. The resulting apple will be lopsided, for the side with defunct seeds will have no hormones to stimulate fruit growth.

Although a heavy frost can severely limit the crop of some apple varieties, others may survive to produce a big crop anyway. For example, following a heavy spring frost in an Illinois orchard, the variety Starkrimson didn't produce its usual commercial-size crop. Golden Delicious and Jonared, however, both yielded normal crops from their second flush of blossoms that escaped the frost.

Fall frosts can also take their toll on apples. If the fruit is exposed to frost before it is ready to pick, the outer tissues may be injured. Then, as the apple continues to mature, this damage will show up on the skin as patches of brown and corky tissue.

The fruit isn't the only part of the tree affected by frost. The growth of foliage can also be slowed down by a freeze, even if there's no visible damage. A study done at Washington State University in 1977 showed that a week after frost the leaves of Golden Delicious trees were respiring at a lower rate than normal. There's no reason to think that this doesn't apply to other apple trees as well, so a cold spring could mean less vegetative growth than normal on your apple trees.

A very severe late spring frost (23° to 24°F) can affect the leaves in a strange way. Leaves exposed to cold like this in late spring will have crinkled margins with burned-looking edges. If you notice leaves with these markings, think back to the spring weather before you try to figure out what disease is attacking your tree!

HARVESTTIME

During the last few weeks before harvest, the cells in the fruit swell with water, the starch is converted into sugar, and the chemical that causes the astringent taste characteristic of unripe apples disappears. Air spaces develop between the cells, which eventually account for 25 percent of the apple's volume. Some varieties have more air spaces than others; those with more air have whiter flesh, while those with less air have flesh with a slightly yellow cast.

As the time of harvest nears, the final color of the outside of the fruit develops. From season to season you may find that the color of the apples you harvest from the same tree varies quite a bit—one year the fruit may be pale, another year it might be more brightly colored. An apple's color is very dependent on the amount of light it receives. Some seasons you may not have pruned quite enough so your apples were shaded; these fruits won't be as highly colored as ones that bask in the sun.

In most cases, apples are ripe when they are easy to pick. However, this isn't a foolproof way to judge ripeness. As the fruit matures, a weak zone develops between the stem and its attachment to the spur. In the variety McIntosh, this zone often forms before the fruit is really at its ripest, so many fruits may fall from the tree before they're at their best eating quality. In contrast, Cortland doesn't come off easily even when the fruit is overripe. To be sure of picking your apples at the height of flavor, pluck one off the tree and sample it.

You should always be cautious when harvesting your apples to avoid damaging the fruit spurs. Lift up on each apple and twist it as you pick so that the stem releases gently from the spur.

STORING THE BOUNTY

Windfalls and apples you accidently drop won't keep as well as unbruised ones, so keep them separate from the unblemished ones and use them up as quickly as you can. If you store your apples, try to keep them as close as possible to the ideal storage conditions we describe in chapter 1 under What to Do with Ripe Fruit. Check your stored apples often, for the saying "One rotten apple spoils the whole barrel" is all too true!

PROBLEMS WITH APPLES

Apples have their share of problems. Some apple diseases and pests are difficult to control, simply because so many people grow apple trees and there are so many trees that aren't given proper care. These can serve as reservoirs for pests and pathogens, making the fight to have clean, worm-free fruit and healthy trees that much harder for the conscientious organic gardener. Fortunately, recent developments in breeding resistant varieties and in creating organic insecticides have made it lots easier to control apple pests. In this section we'll give you a rundown of the most common problems you're likely to encounter and ideas on how to deal with them.

Apples are susceptible to a whole host of fungal diseases, and unfortunately, there are no effective organic controls to fight them. Resistant varieties are the best hope for organic growers, and more of these apple types are appearing on the market. One of the most popular is the excellent variety Liberty. Liberty is a good all-purpose apple that is very resistant to apple scab and cedar apple rust and resistant to fireblight and mildew. In terms of flavor, it's similar to McIntosh.

Fireblight: This is a serious disease of apples, pears and other members of the rose family such as hawthorne, roses, flowering quince and mountain ash. The first official record of it dates from 1780 in the Hudson Valley of New York State. Fireblight is now found throughout the United States and has spread to Europe as well.

This disease is caused by a kind of bacteria that overwinters along the edges of existing cankers (sunken areas in the bark). In early spring, when the weather warms up enough, the bacteria start to grow. In dry climates, insects spread the disease, while in humid areas, rain is responsible. The susceptible parts of the tree are the blossoms, the young vegetative shoots and damaged areas of the bark. In warm climates, trees are infected through the blossoms, but where it is cool at flowering time, the bacteria enter through the young shoots.

Once the bacteria land on a suitable surface, they multiply and then penetrate the plant. An infected blossom looks water-soaked at first, then shrivels and darkens. It looks as if it's been through a fire, hence the name, fireblight. The leaves on an infected tree turn black. Once a tree becomes infected, it's extremely difficult to stop the condition from worsening. The infected tissues ooze bacteria in a thick liquid, which can infect even more of the tree. In the fall, the bacteria form a canker in the bark and overwinter in the tree, where they'll be ready and waiting to afflict more of the plant come next spring. If hail strikes a fireblighted tree, it can really wreak havoc. The wounds made by the hail in the tender tissue open up new sites of infection.

In young trees, a severe infection can occur in the bark, girdling the tree and killing it. Even the roots can succumb. A sucker can act as an entryway for the bacteria, which then travel downward from the infected sucker tip into the main root system, where they can eventually kill the tree.

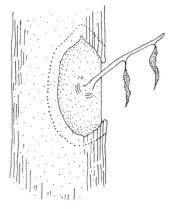

Fireblight Canker: When cankers appear on your trees, remove them in the spring to keep fireblight from spreading. Cut away the canker and an area 2 to 4 inches around it.

As you can tell from these descriptions, a tree with fireblight is not a pretty sight. Although there's really no cure for it, there are some things you can do to prevent fireblight damage in the first place. The most important step you can take is to select a resistant variety. We've already mentioned Liberty, and you can find the names of other varieties in the Guide to Apple Varieties at the end of this chapter. Another preventive measure is to avoid growing your trees in heavy, poorly drained or acid soils. Keep the soil relatively low in nitrogen so that your trees don't grow too fast; this will decrease their susceptible period. For the same reason, avoid heavy pruning that will encourage vigorous new growth.

In the event that fireblight does appear among your trees despite your best efforts, here are a few things you can do to try to stop its spread. If you find small branches infected, cut them off at least 12 inches below any evidence of the disease. Remove the trimmings to the garbage rather than to the compost pile. After each cut, dip your knife or saw in a 10 percent bleach solution so you don't inadvertently spread the bacteria yourself. In springtime, remove any cankers you see by cutting 2 to 4 inches around the area. Be sure to remove any discolored wood along with the bark.

Apple Codling Moth: If you've ever bitten into an apple and encountered a worm, chances are good it was the larva of the apple codling moth—one of the most serious apple pests. This moth lays its eggs on the blossom end of the young apple, and the larvae burrow into the core of the fruit. In the fall, the larvae emerge and find a crack or rough spot in the bark of the tree to hide in and overwinter.

The best weapon against this pest is cleanliness. Don't let wormy windfalls accumulate under your trees—pick them up promptly. You can try to salvage these by cutting up the good parts for sauce. But if you don't want to bother, discard the apples in plastic bags or boxes—just be sure to get them away from your trees so that larvae can't settle in for the winter.

Another strategy is to wrap several layers of corrugated paper around the trunk in the spring and remove it in the fall—you'll find that many of the larvae have spun their cocoons there rather than in the bark. Discard this paper and you've gotten rid of a significant number of pests. If you have only a few trees and you're not squeamish, you can hand-pick the larvae from the bark. Remove any loose bark to deprive the larvae of a good place to overwinter. In the spring, a spraying of *Bacillus thuringiensis* just after the petals drop may catch the larvae as they hatch.

Fruit Tree Leaf Roller: Also called the skeletonizer for the effect it has on the leaves, this is one of the worst apple pests in our area. This caterpillar can totally destroy all of the leaves on a tree in a single season. Luckily for fruit growers you don't have to stand helplessly by while the pest does its dirty work. Leaf rollers are vulnerable to *Bacillus thuringiensis* spray, and the best time to apply it is when the caterpillars are small and living on the undersides of the leaves. When they become larger, they move to the upper surfaces and make themselves protected homes by rolling up the leaves. Once they've reached this stage, the spray has no effect since it can't penetrate the leaves to reach them.

The leaf roller produces two generations in a year. If you had trouble with this pest in the spring, check your trees at dusk in midsummer. Should you spot any moths flying about, examine the leaves in a week or so for the second batch of tiny caterpillars and spray the trees again.

APPLE FRONTIERS

Because of the apple's great commercial importance, research in the areas of rootstocks and resistant varieties is proceeding at a fast clip. A promising development is the series of Michigan Apple Clone (MAC) rootstocks being developed and tested at Michigan State University. MAC-9, also called Mark, is the farthest along and looks like very good news for home apple growers. Mark produces dwarf trees in the size range between trees on M9 and M26 rootstocks. Its roots are strong and hardy and can thrive in a variety of soils. Mark produces a long-lived tree, and most importantly has shown itself to be compatible with all the scion varieties tested so far. Another positive note is that Mark may be available in limited quantities starting in 1985. At least two other rootstocks in this series—MAC-39 and MAC-46—are also potentially excellent dwarfing rootstocks, so we can look forward to more good rootstock choices in the future.

Apple breeding demands great patience, for it takes 20 years or more to develop a new variety to the point where it's ready to introduce on the market. With certain breeding programs, such as the search for better commercial tomatoes, the quality of the crop seems to decline rather than improve. Luckily, this isn't the case with apples. Commercial growers need the same qualities as home growers—hardiness, disease resistance, small but productive trees, early fruiting and good quality apples that store well. You might be surprised to hear that commercial growers value disease resistance, since so many of them use pesticides and fungicides. The truth is, these products are expensive, and they take time and money to apply. If there's a way to raise a marketable crop with a minimum amount of spraying, commercial growers would certainly be interested. As commercial growers push for resistant apple varieties, those of us who garden on a smaller scale will certainly benefit as more of these apples filter down to the home market.

Guide to Apple Varieties

Arkansas Black—Dark red, ripens very late. Zones 5–8. Use fresh, for cooking, sauce. Stores well.

Baldwin—Large, yellow-green with red stripes, sweet, late. Zones 4–7. Use fresh, for cooking, cider, sauce. Stores exceptionally well.

Criterion—Red with yellow, midseason. Zones 5–8. Use fresh or for cooking. Stores well.

Cortland—Large fruit, red with green, midseason. Zones 4–7. Use fresh, for cooking, cider, sauce. Stores well.

Ein Shemer—Medium size, yellow, late. Zones 8–9. Use fresh or for cooking.

Freedom—Red, midseason. Zones 5–8. Resistant to apple scab, cedar apple rust, fireblight; moderately resistant to powdery mildew. Use fresh, for sauce.

Golden Delicious—Small to large, yellow, moderately acid, late. Zones 5–8(9). Use fresh, for cooking, cider, sauce. Stores well.

Granny Smith—Large, green, tart, very late. Zones 6–9. Very susceptible to mildew. Use fresh, for cooking, sauce. Classic baking apple.

Gravenstein—Large, red with yellow, early. Zones 6–8. Use fresh, for cooking, sauce.

Idared—Medium to large, red, late. Zones 5–7. Very susceptible to fireblight; susceptible to mildew. Use fresh, for cooking, sauce. Stores well.

Jerseymac—McIntosh type. Red, moderately acid, early. Zones 4–7. Susceptible to scab and mildew; resistant to fireblight. Use fresh, for cooking, sauce. Stores well.

Jonathan—Medium, red with yellow, tart, late. Zones 5–8. Very susceptible to fireblight; susceptible to mildew. Use fresh, for cooking, cider.

Liberty—Red with yellow, midseason. Zones 4–6. Very resistant to scab and cedar apple rust; resistant to fireblight and mildew. Use fresh, for cooking, sauce.

Lodi—Large, green to yellow, early. Zones 4–8. Very susceptible to fireblight; susceptible to scab; moderately susceptible to mildew. Use for cooking, sauce.

Macoun—Small to medium, red with green, tart, late. Zones 4–7. Use fresh. Stores well.

McIntosh—Medium, red or green, tangy, aromatic flesh, midseason. Zones 4–7. Use fresh, for cooking, cider, sauce.

Mutsu—Large, yellow, very late. Zones 5–8. Use fresh, for cooking, sauce. Stores well.

Northern Spy—Large, red with green, tart, very late. Zones 5–8. Use fresh, for cooking, cider, sauce. Stores well.

Prima—Medium to large, red with yellow, early. Zones 5–8. Very resistant to scab; resistant to fireblight and mildew; susceptible to cedar apple rust. Use for cooking. Stores well.

Priscilla—Large, red with yellow, midseason. Zones 5–8. Very resistant to scab and cedar apple rust; resistant to fireblight and mildew. Use fresh. Stores well.

Red Delicious—Medium to large, red, sweet, late. Zones 5–8. Resistant to fireblight and mildew; very susceptible to scab. Use fresh or for cider. Stores well.

Red Rome—Large, red, tart, late. Zones 5–8. Very susceptible to scab. Use for cooking, sauce. Stores well.

Sir Prize—Yellow, late. Zones 5–8. Very resistant to scab; resistant to mildew; moderately resistant to fireblight and cedar apple rust. Use fresh.

Wealthy—Red, midseason. Zones 4–7. Use fresh, for cooking, cider.

Winesap (Stayman)—Medium, dark red, slightly tart, juicy, very late. Zones 5–8. Resistant to fireblight. Use fresh, for cooking, cider, sauce. Stores well.

Winter Banana—Large, red with yellow, late. Zones 5–8. Use fresh or for cider. Stores well.

Yellow Transparent—Medium, yellow, early. Zones (4)5–8. Use for cooking, sauce.

York Imperial—Medium to large, red with yellow, slightly tart, late. Zones 5–8. Very susceptible to fireblight. Use fresh or for cider. Stores well.

CHAPTER 8

PEARS
AROMATIC AND BUTTERY SOFT

Vital Statistics

FAMILY	**HOW LONG TO BEARING**
Rosaceae	Dwarf 3–5 years
	Standard 6–8 years
SPECIES	
Pyrus communis	**CLIMATE RANGE**
	Zones 5–8 with some varieties in
POLLINATION NEEDS	Zones 4, 9 and 10
Requires cross-pollination	

The buttery, tender-sweet flavor of a well-ripened Bartlett pear in summer seems worlds away from the crisp, crunchy tang of a juicy McIntosh apple in autumn. In truth, they're a lot nearer than you think, for these two fruits are actually very closely related. As you will see throughout the chapter, this kinship results in very similar cultural practices for these fruits.

Pears originated in the same parts of the globe as did apples. Our common garden pear, *Pyrus communis,* came originally from the Near East and Asia Minor and gardeners have been growing it for centuries. Back in 1000 B.C., Homer referred to pears as one of the gifts of the gods, and the Greeks grew named varieties as early as 300 B.C., using grafts and cuttings to propagate them. Some of our present-day varieties were developed in Belgium during the eighteenth century, including such favorites as Anjou, Bosc, Flemish Beauty and Winter Nelis. Bartlett, on the other hand, originated in England in 1796 from a chance seedling.

When most gardeners in North America think of growing pears, they tend to limit themselves to varieties of *Pyrus communis.* This is a shame, for there are other exquisite pear species available, just waiting to be grown. *P. nivalis,* the perry pear, has been grown in parts of England and France for over 400 years and is used to make a special pear wine. Asian or oriental pears (*P. pyrifolia* and *P. ussuriensis*) are becoming more familiar in North American gardens. These tasty fruits look like apples and have the crunchy apple texture, but they carry the distinctive flavor of a pear.

Medley of Pears: There are a number of variations on the basic pear shape. In this grouping, from left to right, is an elongated Bosc, a squat Bartlett, a diminutive Seckel, an apple-shaped Asian pear and a plump Kieffer.

PEAR VARIETIES

While there are hundreds of varieties of pears in existence, only a couple dozen are readily available to the home fruit grower. Pear names can be confusing, for different names may be used for the same variety. For example, Beurre Bosc and Beurre d'Anjou are the same as Bosc and Anjou, respectively. Our Guide to Pear Varieties at the end of the chapter should help straighten things out for you.

Northern growers should be forewarned that because pears are not as hardy as apples, they generally can't be grown as far north. However, we've found that there really aren't any hard-and-fast rules that apply to pear hardiness. For example, experts say that pears in general will suffer a great deal of winter damage when temperatures fall below −20°F. But Dorothy has had no winter damage at all on her Comice and Starking Delicious pears, even after one winter when the mercury plummeted to nearly −30°F. Had she listened to the experts, she wouldn't have planted pears. But she was willing to chance it, and so far she's been lucky. Even if you think you're too far north to grow pears, analyze the microclimate in your yard—with favorable conditions you may be able to succeed.

Growers in the Deep South will be glad to know that pears don't require as much winter chilling as apples, so they can be grown in that part of the country. Some pear varieties require even less than the normal 600 to 900 hours of cold—Twentieth Century and another Asian variety, Shinseiki, are but two examples. Baldwin and Hood are other good low-chill varieties.

Besides climatic considerations, there are other factors to think about when selecting pear varieties. Since pears are especially susceptible to fireblight, you should check to see how much of a problem that disease is in your area. Your local extension agent would be a good person to ask. If fireblight is prevalent, you'll be limited in the pear varieties you can grow, since there are few that show resistance. If fireblight isn't a problem and your climate is a moderate one, you can grow just about any pear variety you like.

Since pears require cross-pollination, you need to grow two varieties unless you're fortunate enough to have a neighbor with a compatible variety within 150 feet. Choosing compatible varieties can get tricky, since some pears can't be used as pollinators. Some of those to watch out for are Magness (the pollen is sterile); Bedford, Bristol Cross and Waite (they don't produce enough pollen); and Beurre d'Amanlis and Pitmaston Duchesse (they'll only pollinate themselves). If you want to grow any of these particular varieties, you'll need to make room for at least two other varieties in your yard. To complicate matters a bit more, some pears, like apples, are mutually incompatible. Bartlett and Seckel won't cross-pollinate, and neither will Louis Bonne and Precoce de Trevoux. The various types of Bartlett pears won't cross-pollinate each other, so you shouldn't plan on growing a Max Red (a red form of Bartlett) and a regular Bartlett without a third tree to serve as a pollinator.

Of course, one of the ground rules is that your pear trees must be blooming at the same time to cross-pollinate. Double-check to make sure the blooming periods of your varieties will overlap. In the case of Bartlett, the weather may disrupt even your best laid plans. Even though Bartlett is technically listed as a midseason bloomer, it does take a while to break winter dormancy. Some years, after an especially mild winter, it may take too long to blossom and miss the blooming time of other pears. So, if the winters in your area are generally mild, you should consider planting a late-blooming variety such as Hardy to ensure pollination of your sluggish Bartlett.

The Asian pears available in America belong to two species. The sand pear, *Pyrus pyrifolia,* is a popular one in China, with several hundred varieties being grown there. Here in North America you have only a few to choose from. Twentieth Century is the most common sand pear variety you'll find listed in nursery catalogs. The fruit is mildly sweet and juicy, but quite grainy. It isn't very hardy, so it can only be grown

Pears with True Grit

Have your ever sunk your teeth into what promised to be the sweet, smooth, juicy flesh of a pear, only to find yourself with a mouthful of what seemed like grit? This unpleasant texture is caused by stone cells within the fruit, which produce thick, hard walls. The more stone cells there are, the grittier the fruit will be. There's nothing you can do to cut down on the number of cells in the fruit, since this trait is inherited rather than caused by environmental conditions. Most popular varieties have few stone cells, if any, which probably helps explain why they're so popular. But some varieties are loaded with these cells.

A group of United States Department of Agriculture scientists rated various varieties on their stone cell content. They used a scale of 1 to 9; 9 meant no stone cells, 5 meant a moderate amount, and 1 meant the pear was chock full of them. The only variety which rated a 9 was Ananas de Courtrai. Bartlett, Gorham, Magness and Moonglow all rated a 7, while Bosc, Conference, Comice, Maxine, Max Red Bartlett and Seckel all scored a 6. Anjou, Clyde, Ewart, Fort Valley and Winter Nelis all rated a 5, while Kieffer, Lincoln, Old Home, Pultney and Waite were grittier, with a rating of 4.

Sometimes a nursery catalog description will mention whether a pear tends to be gritty or not, but you can't always count on it. Your best bet would be to refer to this box before you actually buy a tree to see whether the variety you've chosen was rated, and where it stands. That way you can avoid being disappointed by a gritty harvest.

in mild climate areas in zones 6 through 9. The other oriental pear is commonly called the Eurasian pear (*P. ussuriensis*). This is the hardiest of all cultivated pears and has small, roundish fruit with a rather mild flavor. If you're anxious to grow pears but live where the climate is too harsh for regular kinds, you might consider a hardy Eurasian variety. It won't have the same full-bodied flavor as a Bartlett, but at least a mild pear flavor is better than none!

Satisfying the cross-pollination requirements of Asian pears is a relatively simple matter. There aren't as many mismatches to watch out for as there are with common garden pears. In fact, Asian pears will cross-pollinate with common garden pear varieties.

PEAR ROOTSTOCKS

Since pears are not as important a commercial crop as apples, less research has been done on rootstocks. This is unfortunate, since the rootstocks that are being used today do have some drawbacks and could certainly be improved upon. When you buy a pear tree, chances are it probably is grafted onto Bartlett seedling rootstock — the most commonly used. Because Bartlett seedlings are genetically diverse, they can't be used reliably to produce the fruiting part of the tree, but they are hardy enough to stand up to winter cold. Unfortunately, seedling rootstock is very susceptible to fireblight, pear root aphids and nematodes, as well as to a disease known as pear decline. When properly pruned, trees on seedling rootstock will grow to about 15 feet in height with a spread of up to 25 feet. They'll take anywhere from six to eight years to come into bearing.

Old Home is another variety that's sometimes used as a rootstock. Its number one attribute is resistance to fireblight. Old Home also resists pear decline and will grow better than other pear rootstocks in heavy, wet soils. Old Home is hardy, but it is slow to bear and produces a full-size tree. Small-space gardeners who are looking for an early harvest would be better off staying away from pears on this rootstock.

There are also dwarf pears on the market, which owe their diminutive size to quince rootstocks. The combination of pear scion on quince rootstock isn't new — it's one that's been utilized for centuries in Europe. The reason the graft works is that the quince, *Cydonia oblonga,* is related to the pear. Several of these rootstocks are in use currently. Quince A produces a tree ranging from 30 to 60 percent of standard size, while Quince C results in a half-size tree. Provence quince is the smallest, producing a tree that's one-third standard size. Besides saving space, pears on these rootstocks start to bear within three to five years (as compared with the four to eight years of a standard-size tree). Another plus is that quince rootstocks aren't susceptible to pear decline or pear root aphids.

Despite all the qualities they've got going for them, there are some serious problems with quince rootstocks. First of all, the quince is not as hardy as the pear. Even though the scion may be able to survive in a severe climate, the rootstock may succumb, causing the whole tree to die. In addition, the quince is quite susceptible to fireblight. Also, trees on quince rootstocks have to be staked, since the roots aren't very strongly anchored, and they don't do well in poorly drained soil. Another drawback is that the quince is incompatible with certain pear varieties — the rootstocks produce toxic substances which harm the scion. Fortunately, there's a way to get around this problem — nurserymen use an Old Home interstem between the quince rootstock and the scion in the trees they sell.

The major nurseries, such as Stark Brothers and Miller, all use quince rootstocks to produce their dwarf pear trees. Now that you're aware of the limitations of this rootstock, you can understand why if you live in an area where fireblight is a problem, where winters are cold, or where the soil is heavy, you should forget about growing dwarf pears from the usual sources. This doesn't mean that you must abandon plans for growing dwarf pears altogether. Researchers at Oregon State University have developed a series of rootstocks that are fireblight resistant, hardy, early bearing, well anchored and free of root suckers. These superior rootstocks are crosses of Old Home and Farmingdale and are abbreviated OHxF. Some OHxF clones are dwarfing, while others produce full-size trees. At the present time, the only nursery that uses OHxF rootstocks is Carlton Nursery. (See the Plant Sources Directory at end of the book for their address.)

Oriental pears are grown on several different rootstocks. Bartlett seedling works well, as does sand pear and another oriental species, *Pyrus betulifolia.* Trees on the last rootstock are large and vigorous and they produce an abundant harvest of good-size fruits. You do have to wait a bit longer for this harvest, since these trees come into bearing a year later than other oriental species. *P. betulifolia* rootstock is not susceptible to pear decline, fireblight and pear root aphids, but it isn't as hardy as Bartlett seedling rootstock. Sand pear rootstock, on the other hand, is moderately susceptible to pear decline, very susceptible to woolly pear aphids and mildly resistant to fireblight. In terms of hardiness it lies between *P. betulifolia* and Bartlett seedling. This rootstock produces a standard-size tree.

THE INSIDE STORY ON PEAR FLOWERING AND FRUITING

In general, pears grow much the same way apples do. Like apple trees, pear trees produce their fruit on long-lived spurs, and the fruiting spurs are generally borne on wood at least two years old. As with apples, fruit buds sometimes appear on the ends of one-year-old wood, but they seldom set fruit. If you ever find your pear trees bearing fruit this way, remove the fruit, before the branches bend down under its weight and vegetative growth slows.

Pear flowers are formed the year before they actually bloom. When they do burst forth, they are almost always white, so at a quick glance you might think they bear little resemblance to the often delicate pink apple blossoms. But a closer look will reveal that they both share the same arrangement of five petals. If you compare a blossom of

Springtime Beauty: Pears and other fruit trees give you a double bounty. In the spring you get a visual treat as they cover themselves with clouds of delicate blossoms. And, of course, later in the season you get the pleasure of savoring their luscious fruit.

each, side by side, you'll see that pear blossoms have more abundant stamens with distinctive red anthers. Although pear and apple flowers may differ in subtle ways, a pear will never develop more than ten seeds, just like an apple.

Pears won't blossom until the average daily temperature reaches at least 48°F. They generally bloom a few days ahead of apples, but this early flowering doesn't jeopardize the crop since the blossoms are fairly frost resistant. Most fully opened pear blossoms won't be injured by a temperature of 28°F, although they can be damaged below 26°F. If the flowers have not yet opened, a dip down to 26°F may not harm them.

While a light frost won't kill the flowers, a long cool spell can interfere with pollination. Warm temperatures during the day are essential for bees to carry out their mission of cross-pollination. Besides hampering bee activity, cool weather slows down the pollen tubes. They won't grow fast enough for the eggs to be fertilized before the flowers drop off. This explains why, after a very cool spring, you may not have a good crop, even though the flowers survived just fine. The variety Comice is especially vulnerable to cool spring weather.

Pear trees, like apple trees, are prolific flower producers, so all it takes is a 3 to 5 percent set rate to guarantee you an adequate harvest. Because they have such a low set rate, pears generally don't need to be thinned, but sometimes they can still surprise you with an especially heavy crop. If this happens to your trees, you should thin the pears to one per spur. You won't get as many pears, but it will improve the size of the fruits you do end up harvesting. Most pear trees aren't likely to fall into the habit of biennial bearing, but Winter Nelis can develop this problem. With this variety, you should be especially careful to avoid an overload of fruit. Thin the fruit if you have any concern about the size of the crop.

PEAR SEEDS PLAY A VITAL ROLE

The best quality pears will have at least four seeds. As with apples, the June drop will be greater if there are a lot of fruits with only one seed. Since good pollination is the key to ample seed development, you must plant more than one pear tree to ensure good cross-pollination. And you can't just assume that one pear variety will pollinate another. You need to do your homework to know which varieties are compatible (see Pear Varieties earlier in the chapter).

There are some conditions that can trigger the development of small, seedless fruits. In central California, Bartlett trees are frequently seedless when there haven't been enough pollinators in the vicinity and the trees have ended up being self-pollinated. Seedless Bartlett pears are more common when trees are on seedling rootstocks and when there has been a heavy set of fruit the previous year.

Besides affecting the quality of the pears you harvest this season, seeds can also affect the following year's crop, in a most favorable way. The hormones produced by the seeds which promote fruit growth are stored over the winter within the tree. When the tree breaks dormancy and gets geared up to produce a new crop, these hormones are there, ready to encourage fruit development. This stock of hormones comes in handy, especially when trees face a cold or rainy spring (conditions that work against good pollinatioin, which in turn affects how many seeds develop within each pear). If one year's bumper crop of pears is followed by a cold and miserable spring, there may be just enough of these stored hormones from the large crop to counterbalance the ill effects of the weather. The hormones can be transported back into the fruit, where they'll make up for the absence of an adequate number of seeds.

If you're growing the variety Comice, you'll be interested to know that sometimes the fruit will abort seeds after they've undergone apparently normal development. The longer the seeds grew before being aborted, the larger the fruits, for the seeds will have been producing hormones for a longer time.

PRUNING AND TRAINING PEAR TREES

The first thing you need to know about pruning pear trees is that it's not as clear-cut as it is for apple trees. Some varieties, such as Comice and Anjou, have stiff limbs which do well with the central leader method used for apples. But other varieties, such as Bartlett and Bosc, have more flexible limbs. If these varieties are trained to a central leader, the heavy fruit may force the limbs to bend down to the ground. When that happens the branches stop growing at the tips and instead put their energy into a profusion of water sprouts. These shoot up from the parts of the branches that are close to the horizontal. Since what you're after is a good crop of pears, not water sprouts, open center training is recommended for those pear varieties with flexible limbs. This method is also preferred over the central leader for fireblight-susceptible trees. If one leader in an open center tree becomes infected, it can be pruned away, leaving other productive leaders. In a central leader tree, however, pruning away its single leader can set back the tree seriously.

A few words about any pear pruning are in order. Pears are extremely susceptible to fireblight, and each time you cut away wood, you're exposing tissue, making these trees an even better target for the disease. This doesn't mean you should *never* prune—just make each cut count and try to do the bare minimum of pruning.

CENTRAL LEADER METHOD

When the central leader method is used, pear trees should be pruned and trained as described in chapter 6 under Pruning a Central Leader Tree, with one minor adjustment. Pear trees tend to grow in a more upright fashion, so you must attend to spreading the limbs more persistently. However, they don't need to be spread quite as far as other trees; any angle from 60 to 70 degrees is fine.

Here are some more specific instructions to follow when pruning your pear trees. If you've got an unbranched, one-year-old whip, head it back to 36 inches. You should leave 12 inches between scaffold limbs. Bartlett and Winter Nelis produce lots of water sprouts on their older wood. If your tree develops this problem, you may want to remove some of these suckers in late summer, but only if the tree is very vigorous. A tree that is less vigorous is better off with the suckers on, since a weak tree can succumb very quickly to fireblight.

OPEN CENTER METHOD

You can follow the directions we give in chapter 6 under Pruning an Open Center Tree after you have planted your young tree and headed it back to 20 to 30 inches.

As your tree matures and starts to bear bountiful crops, you may notice that the primary branches begin to sag under the weight of the ripening fruit. If you don't come to the tree's aid, these branches could sag so far that they split off. Help shore up the tree by tying a thick rope (not nylon cord) between the adjacent main branches. To protect the bark, place a layer of thick cloth (old diapers are great) between the tree and the rope. This will put your mind at ease that no damage will occur that could encourage fireblight. You can leave this brace on year-round, but be sure to check it each season to see whether it needs loosening. If you ignore it, it may girdle the branches as they expand.

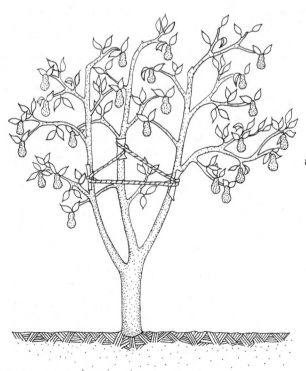

Help for Sagging Branches: Pear trees are notorious for drooping under the weight of a ripening crop. If the branches on your tree start to sag, shore them up by tying a thick rope between the primary scaffolds. Place a thick pad of cloth between the rope and each branch to prevent damage to the bark.

REMEDYING THE SPUR-BOUND TREE

As they get older, pear trees sometimes become spur bound. When this happens, all the growing points become reproductive, so the tree bears smaller and smaller pears and doesn't produce much in the way of new branches and leaves. You can invigorate such trees and get them back on the right track with some careful pruning. Thin out smaller laterals, head back the larger ones, and remove some of the multibranched spurs. This type of painstaking pruning takes time, but it will encourage vegetative growth and in the long run result in a harvest of good-size pears.

Anjou fruit spurs are not as long-lived as those of other pears, so you may want to use these same rejuvenating techniques to renew a Anjou tree. By making the tree more vegetative, you will force it to grow new spurs which will give you a good crop for another four or five years.

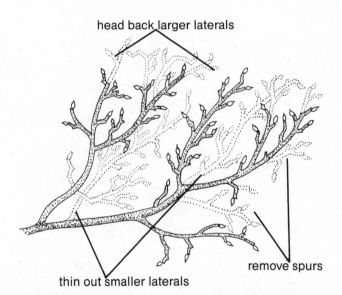

head back larger laterals

thin out smaller laterals

remove spurs

Liberating a Spur-Bound Tree: To encourage larger pears and more vegetative growth, thin out small multibranched laterals, remove multibranched spurs and head back larger laterals to plump, healthy buds.

WHAT PEAR TREES NEED TO GROW

Pears aren't particularly finicky in their cultural needs. In this section we'll highlight the most important things you should remember when tending your pear trees. Like most other fruit trees, pears need a weekly watering session. Without a steady supply of moisture, both fruit and vegetative growth will slow down. Diane really saw what a difference an adequate supply of water can make when she began to grow pears. Although she tried to remember to water her young trees regularly, their location far from the house and garden made it difficult. She's rather embarrassed now to admit it, but they got water only every two or three weeks. Since it was a hot summer and they were planted in rather sandy soil, they should have been watered at least once a week, if not more. Because her pear trees didn't get the water they needed, they grew slowly and put on only 2 to 3 inches of new shoots. Once she put in her trickle irrigation system and her trees started getting a steady supply of water, they really took off, growing almost twice as much as the year before!

Besides water, you should be sure your trees have adequate food for good growth. You'll have to be alert for signs that they're not getting quite enough of a certain nutrient. We give the universal symptoms for various nutrient deficiencies in chapter 1 under The Nutrients Fruiting Plants Need. You'll also find suggestions there on which organic fertilizers to use to get your trees back up to par.

There is one micronutrient that can sometimes pose problems when it's in short supply. Pears need quite a lot of boron, so even before you plant you should check with your local extension agent to see if a deficiency is common in your area. When pears don't get enough boron they suffer from the colorfully named syndrome, blossom blast. The flowers wilt, die, and dry up—but they never quite manage to fall off the tree. They may even cling there until the next year. Blossom blast is more likely to occur during a wet, cool season, and is most likely to afflict trees planted in heavy, poorly drained soils. When it does strike, it usually doesn't affect the entire tree—only certain branches will exhibit the symptoms. Other parts of the tree can also signal a boron deficiency. Leaves with brown, raised spots and leaves that occur in rosettes instead of linearly along the branch are two signs to watch for. If either of these symptoms appears on your trees, spray the blossoms with a boric acid solution (dilute 0.02 pounds [9 grams] of boric acid crystals from a pharmacy in a gallon of water) when they're first opening.

Don't overlook the acid/alkaline balance of the soil your pears are growing in. The soil should have a pH of 6 to 6.5. Trees grown in acid soil are more prone to fireblight, so monitoring your soil's pH is really an important part of disease prevention.

THINGS NO ONE TELLS YOU
ABOUT PEAR RIPENING

If you're in a hurry for a ripe pear, hope for hot weather about 14 to 28 days after the trees are in full bloom. The higher the temperature during this early phase when the cells are dividing, the sooner the fruit will mature. In general, it takes 106 to 124 days after full bloom or when at least half of the flowers are open for the pears to mature.

GOOD PEARS NEED GOOD WEATHER

You can do everything right for your trees throughout the season, but when it comes down to pear maturing and ripening, the weather has more influence than you do. The right temperatures during the last phase of maturation are crucial to properly developed fruit that will ripen nicely. If it's too hot or too cold, the quality of the pears may be seriously affected. That's why so many commercial pear growers are found in northern California and southern Oregon where the climate is moderate.

To give you an idea of just how much the weather can affect pear development, take a look at the variety Anjou. These pears mature and ripen well if the average temperature night and day is between 57° and 62°F. However, if the average temperatures are above 68°F or below 53°F, the fruit won't ripen properly and the harvest will end up being poor in quality. Southern growers should take note: Because of this adverse reaction to warm temperatures, Anjou won't do well very far south.

With Barlett pears, cool temperatures four to five weeks before harvest can cause the fruit to ripen prematurely and fall off the tree. In certain years, commercial crops of Bartletts in Oregon have failed, sometimes completely, because the temperatures were too cool during the crucial weeks when the fruit was maturing. To give you some idea of how critical temperature can be, all it takes to ruin a crop of Bartletts is 48 hours with 44°F nights and 70°F days! A nine-day spell of nights at 50°F and days at 70°F, or three weeks with 44°F nights and 80°F days will have the same effect. There are mitigating circumstances, however. If the nights are warm enough (above 55°F), or the days are hot enough to compensate for the cool nights (90°F), premature ripening won't occur.

From all these temperature combinations you might be led to believe that Bartletts can't be grown reliably in northern or high-altitude areas, but this is apparently not the case. While the causes of premature ripening are not completely understood, it seems that the problem only occurs when temperatures dip down after many days of warm weather. If the days and/or nights have been consistently cool, Bartlett pears seem to ripen at their normal pace.

JUDGING WHEN TO PICK PEARS

Once they're mature, you shouldn't allow pears to ripen on the trees. This may sound a little strange, especially since we extol the virtues of so many other tree-ripened fruits. But tree-ripened pears have a coarse, mealy texture, and their cores become soft, watery and brown. (The exception to this is Asian pears which we'll discuss later in this section.)

At this point, you're probably wondering how you're supposed to know when to pick your pears. There are signs you can watch for. The first signal that pears are entering the maturing stage is a rapid burst of growth to full size. All varieties but Seckel should be at least 2 inches in diameter at their widest part to be considered full size (for Seckel, 1½ inches is a good guide). Varieties such as Bartlett, which become yellow as they ripen, will turn a lighter shade of green when they're mature. The flesh of all varieties will become whiter, and juice will appear on a cut surface. In addition, all varieties except Bosc will separate easily from the spur when the pear is tilted to the side. Once your pears show all the signs of being ready for picking, handle them as if they were fragile eggs even though they may feel hard as a rock. Although they feel firm, pears at this stage will still bruise and form soft spots quite easily.

TIPS ON GETTING PEARS TO RIPEN

After harvesting, your job is to give your pears the conditions that allow them to ripen. Or, if you want to save them for later in the season, you must store them so that they don't ripen prematurely or spoil.

An important factor in getting pears (as well as other fruits) to ripen is the presence of ethylene gas. Cold temperatures act like a switch and turn on the production of ethylene in the fruit. Before some pears will ripen properly, they *must* be placed in cold storage. For example, the varieties Comice, Forelle, Packham's Triumph and Winter Nelis must be stored for a month at 30° to 32°F. Anjou requires two months of cold storage. If you ignore this requirement and try to get your pears to ripen at warmer temperatures, you're not going to have much luck. They will just sit there, hard as rocks, while you become more and more frustrated, wondering why the pears you've looked forward to all season aren't softening. However, there is a way around cold storage, in case you're hard pressed to provide the proper conditions. (You will find a suggestion on how to provide the correct range of cold temperatures in chapter 1 under What to Do with Ripe Fruit.) Place these varieties in a paper bag with ripe pears, apples or other ripe fruits that are giving off ethylene and they will ripen without cold storage.

How Long Pears Keep in Cold Storage

VARIETY (IN ORDER OF RIPENING)	MINIMUM STORAGE TIME (IN MONTHS) 30°–32°F	MAXIMUM STORAGE TIME (IN MONTHS)	
		30°–32°F	40°–42°F
Clapp's Favorite	None	2	None
Bartlett	None	1–1½	½–¾
Seckel	None	3–3½	None
Bosc	None	3–3½	2–2½
Anjou	2	4–6	2–3
Comice	1	2½–3	1½–2
Packham's Triumph	1	5–6	None
Kieffer	None	2½–3	None
Winter Nelis	1	5–7	3–3½
Eldorado	1	4	None

Not all varieties need cold storage to get the ripening process underway. Bartlett, Bosc, Clapp's Favorite and Seckel pears will ripen right off the tree. However, if you want to extend your pear eating season, these varieties may be stored for awhile before being ripened; see the chart How Long Pears Keep in Cold Storage for some idea on how long they'll stay in good shape.

When you're ready to enjoy ripe pears, take them out of cold storage. Place them in a moderately cool location where the temperature is between 60° and 70°F and the relative humidity is as close to 90 percent as possible. In many houses, these conditions can be found in the basement. Spread newspaper on the basement floor and gently place the pears in a single layer. Then, cover with more newspaper so that they don't dry out. Cement floors are especially good for ripening pears, for they are cool and slightly humid. Both of these conditions will help keep the pears from shriveling.

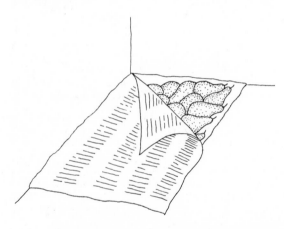

Pears Under Wraps: When you bring pears out of cold storage, put them in a cool place to hasten ripening. A basement with a cement floor is ideal. Spread a single layer of pears on a cushion of newspaper and cover the fruit with more sheets of paper to keep them from drying out.

Getting pears to ripen can be tricky. We've both had the experience of buying pears during the winter, placing them in a bowl, and waiting in vain for them to ripen. Several factors may account for this failure. If the pears are left in cold storage (either in your basement or at a commercial grower's) too long, they often won't ripen at all. However, if the room temperature in your home is too high, pears will fail to soften properly. (For most varieties the point of no return is 85°F, but Kieffer pears won't soften at temperatures above 70°F.)

There are other problems that may beset a ripening batch of pears. At one time or another, we've all cut into what looked like a well-ripened pear only to find that the core was brown and watery. This is the fate of pears that are picked too late. Pears that are picked too early wilt and turn brown on the outside instead of ripening.

Despite all these potential disappointments, there still is nothing quite like a ripe pear, fragrant and soft and dripping with sweet juice. It's fairly easy to tell when a Bartlett pear is ripe; its yellow color and delectable pear aroma are outstanding indicators. Judging other varieties can be trickier, though. The best test for ripeness is to press gently near the stem end of the fruit. If the pear gives to pressure and your fingernail easily penetrates the body of the fruit, it is ripe and ready to eat.

A few words on getting Asian pears to ripen are in order here, since they're not handled the same way as common garden pears. The main difference is that you can let Asian pears ripen right on the tree. Like apples, they are crisp and juicy when ripe. Taste-testing is the best way to judge ripeness. Asian pears store well at 32°F with 90 percent humidity and can stay in good shape as late as Christmas.

PROBLEMS WITH PEARS

Compared with other fruit trees, pears aren't plagued by too many problems. We've highlighted the main ones here.

Fireblight: As we've mentioned throughout the chapter, fireblight is the main problem you're likely to encounter with pears. Should your trees develop this disease, use the methods described in chapter 7 under Problems with Apples to counteract it. If you live in an area where fireblight is prevalent, steer clear of fireblight susceptible varieties and pick only those that show some resistance.

Pear Curl: This is a disease caused by a microorganism that invades the trees. It attacks the phloem and the blockage it causes slows or nearly stops the movement of carbohydrates from the leaves. The visible sign that this is happening is leaves that curl in on themselves. In the fall they'll turn reddish purple instead of the normal autumn yellow and they'll drop prematurely. More serious than early leaf drop is the fact that an advanced case of pear curl can kill your trees.

This disease generally affects young trees. The only cure is a costly one—a visit from a tree surgeon who can administer an injection of antibiotics. (Tree surgeons can be found in the yellow pages, or your local extension agent may be able to recommend one.) If you opt to remove the dying or dead tree, make sure you remove as much of the root mass as you can. When you replant, try to site the new pear tree at least 100 feet from the spot where the sickly pear grew.

Pear Decline: This disease is also caused by a microorganism, perhaps the same one that causes pear curl. The phloem in this case is blocked at the graft union, which causes a slow, lingering death as the tree perishes from starvation. The only cure is the same one that's given for pear curl.

Wind Injury: Those of you raising Anjou pears should be aware that the young leaves are particularly sensitive to wind injury. Scientists ascertained this by placing leaves from several different pear varieties in a wind tunnel. The Anjou leaves were consistently the only ones that showed any damage.

A leaf that's been injured by the wind has blackened edges. If you have an Anjou pear tree that's been exhibiting these symptoms and you can't diagnose any particular pest or disease attack, suspect the wind. Try erecting some sort of windbreak—and should you ever plant another Anjou, give it a sheltered location.

Codling Moth: This insect can be a most annoying pest. See Problems with Apples in chapter 7 for pointers on how to treat an invasion.

Pear Psylla: About the size of an aphid and similar in its behavior, this insect sucks the juice from the leaves, fruit stems and young shoots. Pear psylla can transmit diseases such as pear decline from tree to tree, so it's in your best interest to control the insect. The most effective way to do this is with a dormant oil spray in the spring. Since the psylla hibernates under the bark and emerges when the weather warms up, an early dormant oil treatment will smother it. Once this pest comes out of its winter hideaway, you can also kill it with a soap spray. Luckily, in an orchard where good organic measures are practiced, natural enemies will help to keep pear psylla under control.

PEAR FRONTIERS

It should come as no surprise that because pears are not as important commercially as apples, research on improvements is much less intense. That doesn't mean that no work is being done, however. Breeders are trying to increase resistance to fireblight since that disease is by far the most common problem with pear trees. Rootstock breeding is another priority, but if OHxF rootstocks become more widely used, many of the rootstock problems we now face will be solved.

Breeders are also striving to increase hardiness in some new varieties while decreasing the winter chill requirement for others. Work in these two areas will serve to expand the boundaries of the growing region—good news for northern and southern growers. Good tasting pears which aren't gritty are also being sought by breeders. Unfortunately, all of these positive qualities are hard to combine in a single variety. Most of the fireblight-resistant varieties don't have as good a flavor or texture as a Bartlett, and the low-chill varieties are also often poor in eating quality.

Breeding new varieties of pears is time-consuming, since the tree takes four to eight years to reach bearing age. So, if a new variety of pear comes along which meets your needs, you should silently thank the patient researchers who took the time to develop it.

Guide to Pear Varieties

Anjou—Large, greenish with pink blush, good flavor and texture, late. Zones 5–8. Highly susceptible to fireblight, medium hardy. Use fresh, can, cook.

Baldwin—Tender, good flavor, late. Zones 8–10. Resists fireblight. Can, cook.

Bartlett—Medium to large, sweet, juicy, early to midseason. Zones 5–7. Vigorous but susceptible to fireblight. Use fresh, can, cook.

(continued on next page)

Guide to Pear Varieties
Continued

Bosc—Large, russet, late. Zones 5–8. Susceptible to fireblight. Good keeper, use fresh.

Clapp's Favorite—Large, yellow, good quality and flavor, early. Zones 5–8. Highly susceptible to fireblight, very hardy. Poor keeper, softens quickly after harvest, use fresh.

Comice—Medium-large, juicy, good quality, late. Zones 5–8. Moderately susceptible to fireblight, not too hardy, low chill. Use fresh.

Flemish Beauty—Medium to large, yellow with red blush, good flavor. Zones 4–7. Highly susceptible to fireblight.

Kieffer—Medium to large, yellow, late. Zones 5–9. Resists fireblight, hardy, low chill. For winter storage, can, cook.

Maxine (Starking Delicious)—Large, coarse flesh with good flavor, late. Zones 5–8. Resists fireblight, hardy. Use fresh, can, cook.

Moonglow—Large, yellow, early. Zones 5–8. Resists fireblight, hardy. Pick before dead-ripe, use fresh, can, cook.

Orient—Large, late. Zones 6–8. Resists fireblight, vigorous. Good keeper, can, cook.

Red Bartlett (Max Red)—Large, early to midseason. Zones 5–7. Susceptible to fireblight. Like Bartlett but red. Use fresh, can, cook.

Seckel—Small, very sweet, midseason to late. Zones 5–8. Moderately resistant to fireblight, very hardy. Use fresh, can, cook.

ASIAN VARIETIES

Chojuro—Medium, midseason. Zones 6–8. Susceptible to fireblight. Use fresh.

Shinseiki—Medium, early. Zones 6–9. Susceptible to fireblight. Use fresh.

Twentieth Century—Medium, midseason. Zones 6–9. Susceptible to fireblight. Use fresh.

CHAPTER 9

CHERRIES
SWEET AND SOUR BITE-SIZE TREATS

Vital Statistics

FAMILY Rosaceae	**HOW LONG TO BEARING** Sweet cherry 3–6 years Sour cherry 3–5 years
SPECIES *Prunus avium* (sweet cherry) *P. cerasus* (sour cherry)	**CLIMATE RANGE** Sweet cherry Zones 5–8 Sour cherry Zones 4–8
POLLINATION NEEDS Sweet cherries require cross-pollination; not all varieties are compatible pollinators.	

If you've never plucked a ripe sweet cherry straight from the tree and popped it into your mouth, you don't know just how delicious cherries can really be. Cherries are very perishable; they begin to lose their luscious flavor and juicy firmness as soon as they are picked. The only way to experience cherries at their best is to grow your own. In fact, to some people, the importance of true cherry flavor is an overriding consideration in life. One of the first reasons Diane's fiance (now husband) gave when asked why he wanted to marry her was, "Because the family has cherry trees!" What better dowry could one ask for?

Now that we've told you what your taste buds are missing if you don't grow your own cherries, we need to point out that they are an especially demanding crop. Cherries won't grow in much of the country, for most varieties can't take too much cold, and they also don't do well where winters remain too warm. While some cherries can be raised in the northern tier of states, the Carolinas in the East and central California in the West are about the southernmost limits for cherry growing. Most cherries are only available as standard-size trees which grow to an impressive height, even when properly pruned. To complicate matters even further for small-space gardeners, these big trees require another cherry of a different variety within 40 feet for proper pollination.

The news about cherries isn't all bad, though. In the last few years, a few sweet cherry varieties have become available in "compact" form—genetic dwarfs that grow anywhere from 8 to 14 feet tall. Two of these compact varieties, Compact Stella and Starkrimson, are also self-pollinating. So, if you have a small yard and want to grow cherries, you need not despair!

While we've been talking primarily about sweet cherries here, pie cherries are another homegrown delight (some people know them as sour cherries). These cherries are almost impossible to obtain unless you grow them yourself or have a generous friend. Fortunately for home gardeners, pie cherry trees are smaller than sweet cherries, and they're hardier and self-pollinating. Pie cherries can also be grown successfully further south than sweet cherries.

TRACING THE CHERRY'S ORIGIN

Cherries are native to the area extending from the Caspian and Black Seas east to northern India and China. Like pears, cherries were being grown in Greece by the year 300 B.C. and may well have been cultivated before then for their beautiful wood rather than the modest-size fruit. The Romans followed the lead of the Greeks in growing cherries, but with the fall of the Roman Empire, cherry cultivation apparently ceased until the seventeenth century.

Among their assorted baggage, European settlers brought cherries to the New World. The most important varieties we grow today—Bing, Lambert and Republican—were all selected by the Lewelling brothers, who were avid amateur horticulturists. They carefully transported 300 cherry trees west to Milwaukie, Oregon, by oxcart in 1847. Their task of selecting was eased by the fact that cherries, unlike most fruits, will often breed true from seed.

CHERRY VARIETIES

To help you sort through all the tempting cherry varieties you see in catalogs, we've highlighted some of the more noteworthy ones in the Guide to Cherry Varieties at the end of the chapter. Before we give you some general pointers here on how to go about selecting varieties to grow, we'll explain the basic difference between sweet and sour cherries. Sour cherries (*Prunus cerasus*) have twice the number of chromosomes as their sweeter cousins (*P. avium*) and are thought to have arisen from a hybrid between sweet cherries and another cherry species. If you see mention of the Duke cherry (*P. gondouini*) in catalogs and wonder how it's related to other cherries, here's your answer. It is intermediate between sweet and sour cherries in taste, size and hardiness because it is a hybrid between those two types.

WHICH SWEET CHERRY IS FOR YOU?

Choosing which sweet cherries to grow can be tricky. Not only must you be sure to plant varieties that will cross-pollinate, but you must select trees that will grow well in your area. Most cherry varieties thrive only in zones 5 to 7; if you live either north or south of this range, your choice in cherries is quite limited.

It's unfortunate that three especially popular varieties—Lambert, Bing and Royal Ann (also known as Napoleon)—can't pollinate one another. Diane learned this the hard way by planting Lambert, Rainier and Royal Ann in her yard. After she had waited patiently several years for her trees to bear fruit, they finally burst into blossom. First came the Royal Ann, laden with lovely white flowers. Before that tree finished blooming, the Lambert started in, with Ranier blossoming forth while Lambert was still in flower. Diane was looking forward to a large and varied crop, but she ended up with not a single Royal Ann cherry. The Lambert and Ranier cross-pollinated, but Lambert couldn't pollinate the Royal Ann, and Ranier bloomed too late to do so.

The closest you can come to carefree pollination is with a cherry variety like Republican which can cross with any other varieties. There are also a few cherries—notably Starkrimson and Stella—that can self-pollinate. Unless you choose one of these varieties, you're going to have to pay special attention to bloom times and compatibility. Because the cherry flower is only receptive for a short period, choosing varieties that bloom at the same time is especially critical, as Diane found out. If you select varieties which will cross-pollinate and which bear fruit at about the same time, their periods of bloom should overlap sufficiently to produce a good crop. But other factors can influence pollination, too. For example, if Lambert is planted in the warmer parts of California, it may not amass enough chilling hours in the winter. That will lead it to break dormancy late, after all its potential pollinators have finished blooming.

We've tried to make choosing sweet cherry varieties easier by drawing up a chart that covers the pollination habits of all the sweet cherries we discuss in the Guide to Cherry Varieties. See Matchmaking: Finding Pollinators for Sweet Cherries for guidance.

If you have a small yard but want to grow Royal Ann, Bing or any other variety that only produces full-size trees, you could plant Compact Stella or Starkrimson as a pollinator. That way, you can accommodate two trees in less space than two full-size varieties would take. Should you be lucky enough to have a neighbor

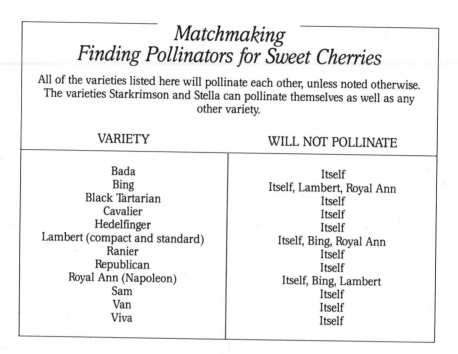

Matchmaking
Finding Pollinators for Sweet Cherries

All of the varieties listed here will pollinate each other, unless noted otherwise.
The varieties Starkrimson and Stella can pollinate themselves as well as any
other variety.

VARIETY	WILL NOT POLLINATE
Bada	Itself
Bing	Itself, Lambert, Royal Ann
Black Tartarian	Itself
Cavalier	Itself
Hedelfinger	Itself
Lambert (compact and standard)	Itself, Bing, Royal Ann
Ranier	Itself
Republican	Itself
Royal Ann (Napoleon)	Itself, Bing, Lambert
Sam	Itself
Van	Itself
Viva	Itself

with some cherry trees within 40 feet of yours, find out which varieties they
are and plant one of your own which will cross-pollinate.

SELECTING A SOUR CHERRY TREE

Sour cherry trees are easier to select than sweet ones. For one thing, there are fewer
varieties to choose from! In addition, all sour cherries are self-pollinating, so you don't
have to worry about providing compatible varieties for cross-pollination. (You should
also know that sweet and sour cherries won't cross-pollinate.) Since you only need to
plant one tree, you can still get a good crop in a minimum of space.

CHERRY ROOTSTOCKS

As is the case with other tree fruits, cherry scions are grafted onto rootstocks to
produce the trees you buy. But unlike other fruits, particularly apples, for which
many different rootstocks are used in grafting nursery trees, only two types of
rootstocks are commonly used for both sweet and sour cherries. One of them,
Mazzard (*Prunus avium*), is a sweet cherry that grows wild in the eastern United
States. The tree gets as tall as 40 feet and produces small, dry fruit. Mazzard is the

most common cherry rootstock. Trees grown on these roots are large and vigorous and somewhat slow to bear. One big plus is that they're tolerant of wet soils.

The other common cherry rootstock is Mahaleb (*Prunus mahaleb*). Mahaleb trees are two-thirds to three-quarters the size of those on Mazzard rootstock, and they come into bearing earlier. However, Mahaleb roots are apparently much tastier to gophers than Mazzard, and Mahaleb doesn't do as well in soggy soil. It's also susceptible to root rot fungi, though more resistant to crown gall than Mazzard. The reason it's so widely used, despite all its less than desirable traits, is that it's convenient for nurserymen to grow. At this time, all cherry trees from Stark Brothers are grafted onto Mahaleb rootstock.

Even though these two rootstocks are the only ones used currently, they may not always have a corner on the market. Colt is a new cherry rootstock developed at the East Malling Agricultural Station in England that is already being tested commercially and may soon make its way into the home fruit tree market. It's desirable because it reduces tree size by 14 to 20 percent, but a major drawback is that it's not cold hardy. In addition, trees on Colt rootstock are slow to drop their leaves in the fall and reluctant to enter winter dormancy. So, if you live in an area with cold or early winters, this probably wouldn't be the best rootstock for your trees.

CHERRY BLOSSOMS

Those clouds of white blossoms that unfurl each spring can almost take your breath away. But don't be fooled by the delicate appearance. Cherry blossoms aren't only for show — they're there to be pollinated so that they can develop into plump, juicy fruit. What a lovely bonus — these trees grace your yard with their beauty and then go on to provide you with delectable fruit. To help you understand the transformation from flower to fruit, here's a description of the flowering process all cherry trees go through.

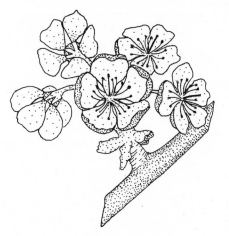

Clouds of White: In the spring, your yard will be graced by the presence of cherry trees in bloom, covered with clusters of white flowers like the ones shown here.

WHERE THE FLOWERS COME FROM

Like apple and pear trees, cherry trees bear their flowers on spurs. But there are important differences between apple and cherry spurs. Apples have mixed buds which produce both leaves and flowers, but cherry flower buds are separate from the leaf buds. At the tip of a cherry spur you'll find a leaf bud, with flower buds along the sides. In the spring, the flower buds open first, followed by the leaf buds. Cherries ripen much earlier than other tree fruits, generally in July, and only after that do the flowers for the next year's crop enter the early stages of development, along with the leaf buds. Because there is a vegetative bud at the very end, a cherry spur will continue to grow straight, unlike the apple spur which has a zigzag growth pattern and can sometimes become branched.

We've been talking about sweet cherries, but basically everything we've said applies to sour cherries as well. Sour cherries undergo the same flowering and fruiting cycle, but they flower, ripen and initiate flowers a bit later in the season.

If the weather is hot during the summer when the flower buds for next season are being formed, they may develop two pistils instead of the usual one. You'll be able to

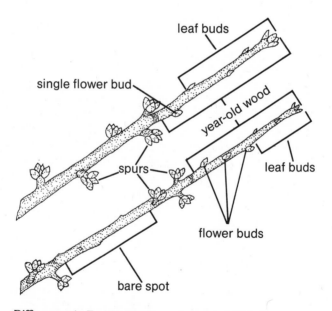

leaf buds

single flower bud

year-old wood

leaf buds

spurs

flower buds

bare spot

Differences in Fruiting Patterns: Sweet and sour cherries both bear fruit on spurs and year-old wood. The difference is *where* the fruit appears on the wood. On a sweet cherry, shown at top, a single flower will appear at the base of the new wood. On a sour cherry, flowers appear along more of the year-old wood. This isn't necessarily an advantage. In future years, that stretch of wood won't develop any leaf or fruit buds, so you'll be left with an unproductive expanse of branch. Careful pruning can compensate for this fruiting pattern in sour cherries.

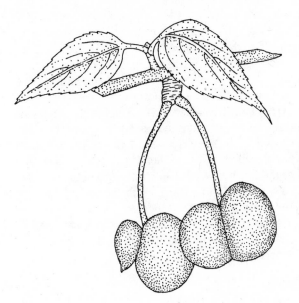

Oddball Cherries: The weather, not any pest or disease, is responsible for the appearance of double and spurred cherries on your tree. Even though they may look a little strange, they're still good to eat.

tell which flowers this happened to because they will develop into either double or spurred cherries. In Oregon, Diane's family seldom had this difficulty with their crop because the summers were not hot. But farther south, in northern California, double and spurred cherries are sometimes a major problem for commercial growers. Since the flavor of the fruit isn't affected, doubling and spurring needn't bother home gardeners.

In addition to the flower buds on spurs, sweet cherries may develop flowers on one-year-old wood, but usually just one bud at the very base. Sour cherries, in contrast, are more prolific. They may develop lateral flower buds for a length of 6 inches or more along the one-year-old wood. If this happens to your sour cherry trees, you may be delighted, imagining your freezer full of cherries ready to bake into pies during the winter. We hate to ruin that wonderful image, but you really shouldn't be overjoyed. The wood that's bearing flowers one year won't develop buds of any kind in subsequent years; it will just remain bare. Over time, your trees will have more and more bare spots and may produce too few leaves to support good overall growth. As we'll see later, there are pruning techniques that you can use to help encourage vegetative growth on your sour cherry trees. With proper pruning, more lateral leaf buds will be produced. These leaves in turn will provide nourishment for a larger crop and they'll serve as points from which more fruit spurs can develop. Unlike the one-shot lateral buds, sour cherry fruit spurs will flower vigorously for two or three years and less so thereafter.

Flower buds on spurs have another advantage over lateral ones—they are hardier. This is probably because the wood they appear on is older and therefore has been able to store more carbohydrates, which helps protect them from winter's cold.

SPURS NEED TLC

Sweet cherry spurs last much longer than those of sour cherries—10 to 12 years. You must be especially careful not to damage or rip off the spurs when you pick sweet cherries. Diane had this fact impressed upon her very early in life. As a child she was warned over and over to be very careful when she picked her grandmother's cherries. This was easier said than done, for the family was harvesting Royal Ann cherries for the maraschino cherry trade. Because they were destined for this commercial use, they had to be picked before they were completely ripe and had reached full color. (Cherries used for maraschinos are bleached, and pale cherries bleach more readily than those with a red blush.) Since the cherries weren't fully ripe, they didn't separate from the spurs easily. And because maraschino cherries with the stems are more desirable than those without, Diane learned how to coax the cherry and stem to come off together, all without damaging the spur. Many cherry orchardists won't allow customers to pick their own cherries for this very reason—they're afraid the spurs will be damaged by careless, inexperienced pickers.

A FEW WORDS
ON POLLINATION AND FRUIT SET

Even when you've selected perfectly compatible pollinators, pollination isn't guaranteed. Sweet cherry trees flower early, about two weeks before apples, so late spring frosts may ruin your chances for an abundant harvest. To further complicate matters, all cherry flowers are only receptive to pollen for a short time. It's best if they're pollinated on the same day they open; by the second day, the flowers are already going downhill. At the end of a week's time, any flowers that haven't been successfully pollinated never will be. If your trees start to flower during cool, damp weather that stays that way for four or more days, don't expect to harvest very many cherries. Like other fruits, cherries depend on bees for pollination, and when the weather turns bad, the bees aren't likely to be out visiting your trees. Because cherries are so small, they need a set rate of from 25 to 50 percent to produce a good harvest. This is high, so good weather and plenty of bees are especially critical at cherry blossom time. If the spring is usually wet in your area, be sure to choose a variety that tolerates wet weather or better yet try choosing a self-pollinating variety. When the bees don't have to travel from one tree to another, the pollination rate is much higher.

Each cherry flower has two egg cells, but only one of these is fertilized, so there's just one stony pit in the center of the fruit. After the egg cell has been fertilized, the fruit begins to grow. Cherries continue to increase in size right up to maturity and sometimes beyond. An interesting study of Bing cherries conducted in 1981 in Prosser, Washington, showed that when the crop was light, the fruit reached full size even before it took on its bright red color. With a heavy crop, the fruit slowly kept expanding but never quite reached full size, even though it was left on the tree until it was overmature and had begun to shrivel. What this demonstrates is that a large crop can slow the rate at which all the fruit increases in size. You may have more cherries, but they're likely to be smaller at picking time than fruit from a lightly bearing tree.

This may lead you to believe that cherries need to be thinned—but in reality they're usually not thinned like the large tree fruits. It's simply too much work to go over trees covered with so much tiny fruit. However, if you're concerned that your trees are carrying too many cherries, your best course of action is just to keep them well watered and well nourished to help them produce the largest fruit possible.

SOME BASIC CHERRY PLANTING ADVICE

To preserve the life of your cherry trees, be sure to plant them in a well-drained spot, for these trees languish in soggy soils. Cherry roots, along with those of apricots and peaches, release hydrogen cyanide when surrounded by waterlogged ground. This phenomenon is discussed more in detail under Getting Peach Trees Off to a Good Start in chapter 11.

While some fruit trees need to be planted slightly deeper than they grew in the nursery, this is not true for cherries. Cherries are very sensitive to crown rot fungi, which may attack if you plant the trees too deeply. Set the trees in the ground at the same depth they grew in the nursery—but no deeper.

Cherry trees do best in rich soils, preferably where the topsoil is 4 feet deep. Trees grown in shallow soil will come into bearing before those in deep soil, but their overall production will never match that of trees with plenty of rich soil around their roots.

The modest little tree you buy, unless it is a compact variety or a sour cherry, will grow to be a thick limbed giant. Even though you might be tempted to skimp on space, do yourself a favor and give your cherry tree plenty of room right from the start. Commercial orchardists sometimes plant sweet cherries as close as 20 feet apart and then prune them back every year. But for the home grower, it's much better to allow a full-size tree a circle 25 to 30 feet in diameter. Otherwise, it will outgrow its space in 10 to 15 years. Sour cherries don't need quite as much space—they can do quite nicely given a circle with a 20-foot diameter to grow in. The dwarf or compact trees of either kind of cherry need only about 8 to 10 feet of growing room.

TIPS ON PRUNING SWEET CHERRY TREES

Because they get so large, you should train standard-size sweet cherry trees to an open center shape. (See Pruning an Open Center Tree in chapter 6 for directions.) This encourages them to grow outward instead of upward, which in turn controls their height so they're easier to pick and allows sunlight to reach their centers. Get your new trees off to a good start by heading them back to 30 to 36 inches.

There's a direct correlation between sunshine and good-quality fruit. When Diane was a little girl, she would gaze up to the tops of the huge old trees in the family orchard and see the abundant, bright fruit waiting there to be picked. She wondered why most of the cherries and particularly the brightest, plumpest ones were always the farthest away and the hardest to pick. Now she knows there's a reason for this, and it's not just the perversity of nature. Those old trees had become so crowded that the sun wasn't able to reach the lower branches. Because the top branches got the most sunlight they produced the most fruit buds. And, thanks to all those rays of sunlight, the uppermost cherries were the first to ripen and turn a tempting red.

With sweet cherries, the key to pruning is to do it consistently. If you don't prune your trees each year, they will quickly get away from you and grow too tall for convenient picking. A friend of ours who inherited an orchard suffers from the philosophy of her father, who started the orchard. He didn't like to prune the trees because he felt they should be free, like his own spirit. That's a lovely philosophical idea, but a practical disaster for a cherry orchard! Now those same trees, which were free for too long, have had their tops radically sawed off, and they are still difficult to pick from.

If you're growing compact sweet cherries, train them to a central leader shape just like apples (see Pruning a Central Leader Tree in chapter 6 for guidelines on how to do this). With this method you'll enhance their space-saving nature by keeping them from taking up as much room horizontally. Head back your newly planted trees to 30 or 34 inches. Choose laterals for scaffolds that are at least 8 inches apart along the trunk. One way cherries differ from apples is that they won't produce laterals from the trunk very far below the tip of the leader. For this reason, you should head back the leader each year to the height of the next desirable lateral. For example, a tree with laterals at 36, 42 and 50 inches and a leader 75 inches tall will end up with a stretch of trunk 15 inches or longer without any laterals if the leader isn't headed back. Once you head the leader back to 58 inches, however, you'll get properly spaced laterals.

TIPS ON PRUNING SOUR CHERRY TREES

With sour cherries, you have a choice. You can train them either to an open center or to a central leader. A central leader tree takes up less space from side to side but requires more pruning attention from you. The open center tree may take up more room, but it demands less of your time as far as pruning is concerned. And it also gives you several strong limbs to rest a ladder on (a feature that can be especially helpful with the larger sour cherry trees).

Consider the merits of both shapes. If you want a central leader tree, head the young tree back to 30 or 34 inches; then follow the general directions given in chapter 6. For an open center tree, follow the basic instructions given in the same chapter, after heading the newly planted tree back to 26 or 30 inches. Be sure to leave more than two scaffold limbs; a sour cherry with only two scaffolds will tend to develop into a wider tree that will take up more room. Choose scaffold limbs about 6 inches apart around the trunk with wide crotch angles. How you prune out limbs which interfere with the scaffolds can affect how those scaffold limbs grow. If you need to remove a lateral above one of your chosen scaffolds, leave a 4- to 6-inch stub when you cut. Trimming the lateral normally will cause the scaffold below to grow more vertically, as if to fill in the space left by the missing limb. But if you leave the stub, the scaffold grows as if the entire limb were still in place! The next year, you can go back and cut the stub.

Don't forget that pie cherry trees are the one kind of fruit tree in which you may need to encourage vegetative growth to keep from ending up with a lot of bare wood. Dr. P. C. Crandall found a way to influence Montmorency trees to produce roughly a fifty-fifty mix of lateral leaf buds and flower buds on new growth instead of the usual preponderance of flower buds. He discovered he could increase the number of lateral leaf buds by trimming back the ends of the branches an inch or two in June. Apparently, cutting back sour cherries before the buds are formed somehow stimulates more vegetative buds to develop. So, if your sour cherries tend to produce many lateral flower buds, leaving bare branches in their wake, follow Dr. Crandall's lead. You should also make sure the trees have plenty of nitrogen to further encourage vegetative growth.

FOOD FOR GROWTH

You can plant and prune your cherry trees by the book, but if you don't give them what they need to grow, they're going to disappoint you. Nitrogen is the principal nutrient cherries require for good growth. An ample supply increases the number of fruit buds per spur and encourages longer shoots to grow. Nitrogen can also invigorate old spurs and increase yield.

Starting when the trees are about two years old, you should give them 0.05 pounds of nitrogen per tree for each year of age in the spring. (See The Nutrients Fruiting Plants Need in chapter 1 for tips on good nitrogen sources, as well as sources for the other nutrients we discuss in this section.)

One way to gauge whether your trees are getting enough nitrogen is to monitor their growth rate. Before they begin to bear, standard sweet cherry branches should grow from 22 to 36 inches a year. However, the branches of a bearing sweet cherry should grow only 8 to 12 inches each year. Full-size sour cherry trees should put on 12 to 24 inches the first few years, then slow down to 6 to 12 inches after they begin to bear. Once sour cherry trees reach full size, lateral branches should grow at least 6 inches annually. This is necessary to promote the production of lateral leaf buds and to provide space for new spurs to develop, since the spurs are vigorous for only a few years. If your cherry trees' growth is slow and the leaves are yellowish, you should suspect a nitrogen deficiency.

Other nutrient deficiencies are rarely a problem with sweet cherries, probably because their root systems are so extensive. But this doesn't mean that they *never* crop up. If sweet cherries need phosphorus, they show the universal symptoms we describe in chapter 1. Sour cherry trees low on potassium will have leaves that curl upward, especially on spurs and the ends of branches. If the problem isn't corrected, the leaves may become bronzed on the upper surface, and the edges may blacken and die. In both sweet and sour cherries, a potassium deficiency can result in smaller fruit. A few words of caution are necessary here: When you're adding potassium to counterbalance a deficiency, don't go overboard. Too much potassium can be just as bad as too little. Sour cherries from a tree with an excess of potassium become soft and lose too much juice when pitted.

If your trees are bearing hard, shriveled cherries, that's the telltale sign of a boron deficiency. Apply ¼ to ½ pound of household borax to the soil under a large cherry tree (20 feet or taller) and less than ¼ pound if the tree is 10 feet or smaller. Use a judicious eye when you measure the borax; too much boron can damage the tree.

If the leaves at the tips of the shoots are crinkly or yellow, your trees probably suffer from a zinc deficiency. When the deficiency is really serious, the leaves will form in rosettes, and they'll be on the small side. Usually, trees mulched with manure and/or hay won't have a problem with too little zinc, since those are both good sources. If your trees aren't mulched and symptoms appear, that should be incentive enough for you to mulch them.

Although it's technically not a food, water is just as important as nutrients in promoting good growth and high-quality harvests. Cherries are thirsty trees that thrive on plenty of water (but that doesn't mean they like soggy soil—we've already pointed out the dangers of "drowning" these trees). Since the fruit is maturing at the same time vegetative growth is taking place, cherries need lots of water throughout the spring and up until harvest.

In the 1970s, during a drought in eastern Washington, horticulturists studied cherry trees to see to what degree water stress affected them. Commercial growers wanted to know just how much water they needed to truck in to keep their trees alive. The researchers found that water stress had caused some outer limbs to die and had reduced the size of the fruit, but the number of flower buds initiated for the next year was not affected. This is rather unusual for a fruit tree—under water stress most trees won't put out many buds. Even though there were a good number of buds, the fruit that ultimately formed from them was undersize. It took the trees three years to recover in terms of producing full-size cherries from just one year of water stress. This should impress upon you how important water is to these trees.

HARVESTING AND STORING CHERRIES

For the fullest flavored and juiciest cherries, you should pick them at the peak of ripeness. Many people think of ripe cherries as being an unmistakably bright cherry red. But there's more than one hue that can signify a perfectly ripe cherry, depending on the variety. For example, ripe Lamberts are a dark mahogany, while Royal Anns are yellow with a red blush. Pie cherries are either bright cherry red or dark red, depending once again on the variety. Your ability to use color as a guide to ripeness will develop over several seasons of experience with your own particular varieties.

Color aside, sweet cherries are easy to test for ripeness—just pluck one carefully and taste it. To pick sweet cherries, pull gently on each stem as you twist it upward. If

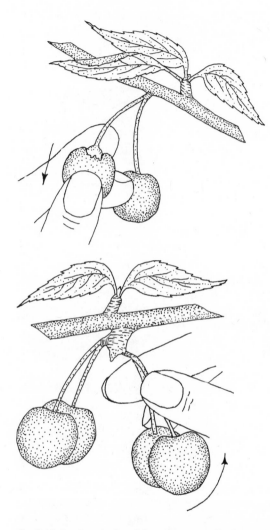

Cherry Harvest: Sour cherries, shown at top, are ready to harvest when the fruit separates easily from the pit (which remains on the stem). Taste sweet cherries to judge when they're ripe. To harvest, pull gently on their stems as you twist them upwards. Always be careful not to damage the spurs.

the fruit is ripe, it should separate easily from the spur. Cherries seem to bruise at the slightest touch, so don't just toss them into a bucket as you pick; try to be gentle. Don't let the fruit sit in the sun while you're continuing to harvest; black cherries in particular will absorb heat very quickly. This raises their internal temperature and provides the right environment for bacterial or fungal disease to multiply rapidly, which means your cherries are likely to spoil faster. If you can't cool the fruit right away, at least set it in a shady spot until it can be refrigerated.

Sweet cherries are quite perishable, so you should make a point to eat or preserve them as soon as possible after picking. Their small size allows them to loose water rapidly. Try to cull out any split or bruised cherries right away. If any of these damaged cherries are buried in a box, they can quickly rot and spread their blight to the other fruit. The standard recommendation you'll hear is to store cherries with the stems left on. We used to go along with that, until we heard some evidence to the contrary. A cherry grower here in Montana recently sent his daughter in Alaska two boxes of sweet cherries, one with stems and one without. To everyone's surprise, the cherries without the stems were in much better condition upon arrival in the Far North than were those with stems intact!

If you must store some of your fresh sweet cherries for a few days, at least give them the best possible conditions. We explain what these are under What to Do with Ripe Fruit in chapter 1. We've found that for storing small quantities of cherries in the refrigerator, a celery crisper (Tupperware makes one) is excellent. This gadget is an airtight box with a plastic screen that rests on the bottom. You put a small quantity of water in the box and place your cherries on the screen, which keeps the fruit from touching the liquid. The crisper holds about 3 pounds of cherries and will keep them fresh up to two weeks.

If you've got your heart set on canning some sweet cherries to see you through the winter, try to get it done within a day of picking. The less time you let the cherries stand, the firmer the texture and the richer the flavor the canned fruit will have. (See the Guide to Cherry Varieties for the names of sweet cherries that can well.)

We haven't forgotten those of you who are growing pie cherries. It's a cinch to tell when your cherries are ready to pick—just pull downward on the fruit. If the pit stays attached to the stem as the fruit slides off, the cherries are ready. Pie cherries are quite soft and are very messy to pick. In fact, it's just about impossible not to get thoroughly sticky with juice before you're through. Once you've picked the cherries, you should walk right into your kitchen and bake a pie or start canning or freezing immediately. It's even more important to deal with pie cherries the same day you pick them, since they're so soft and deteriorate almost before your eyes. Freezing is much easier than canning, and the cherries retain much more of their flavor, texture and color.

If you pit a lot of cherries, you should be kind to yourself and buy a good pitter. An especially efficient model we've seen has a little tray on which you can place about a dozen cherries. One at a time, the cherries roll down onto a washer with a hole in it. You push a plunger down through the cherry, the pit falls into a container below and, as you raise the plunger, the pitted cherry rolls out along a funnel-shaped exit. At the same time another cherry is set into motion to roll onto the washer. With this gadget, you can pit about two cherries per second, without wearing out your hands or mangling the fruit.

PROBLEMS WITH CHERRIES

The most prevalent disease you should be on the lookout for in cherry trees is bacterial gummosis—especially in wet climates. When this disease strikes, long, gummy, damp-looking patches appear along the trunk and/or branches. As the disease progresses, it eventually girdles parts of the tree. At this time, there's nothing you can do to cure a tree once it comes down with gummosis. Because they're particularly susceptible don't plant the varieties Bing, Lambert, Napoleon and Van in wet areas. Corum and Sam are two resistant varieties to look for.

The main cherry pest is the cherry fruit fly, which is a problem across the country. It looks just like the tiny fruit flies you sometimes find hovering around your ripe bananas or bowls of fruit on a countertop.

This particular fruit fly lays its eggs inside the cherry while the fruit is still green, and the larvae develop near the pit. You're likely to get a bit of protein along with your fruit when you bit into an infested cherry. If the flies have laid their eggs late, the larvae will be small and barely noticeable. But if they were able to attack early, you will find fat, white grubs in the fruit.

In states where cherries are grown commercially, such as Oregon, Michigan and Montana, growers (and that includes people with cherry trees in their backyards) are required by law to combat this pest with pesticide sprays. If you live in one of these states and want to grow cherries, you should phone your local extension agent for details about the side-effects of the recommended pesticides. Some of them are devastating to bees and other beneficial insects and may persist for a long period of time, while others are less toxic and more short-lived.

In states where spraying is not mandatory, you may be able to control cherry fruit flies somewhat by hanging sticky homemade traps in your trees. It seems that these pests are partial to yellow, so you'll get good results if you use old yellow tennis balls or small boards painted yellow as the bases for the sticky substance. Coat them with some SAE motor oil or heavy mineral oil and suspend them from the branches of your trees. The motor oil will remain effective for two to three weeks, and the mineral oil will be good for up to two weeks. Once they lose their stickiness you'll have to repeat the process. You can clean off the boards and reapply the oil, but you're better off discarding the used tennis balls.

You can also try spraying with rotenone once you notice the flies are out. You'll have to spray often throughout the season, since rotenone washes off the leaves quickly and loses its effectiveness rapidly as well.

Although we have mentioned throughout the book that most fruit trees make good lawn trees, we won't make the same claim for cherries. Diane has many painful childhood memories of walking barefoot across the grass, only to be stung by a hornet that was sucking the juice from a fallen cherry. Her dog, however, gained a bonanza in rotting, fermenting fruit and would gorge herself until she began to stagger a bit, slightly tipsy from her gluttony.

It's almost impossible to avoid the problem of messy, rotting, insect-infested fruit under a cherry tree, for it's very difficult to pick every last cherry from a mature tree. Almost inevitably, quite a few end up on the ground. If your tree is planted in a mulched orchard or in bare ground, it isn't hard to keep the area clean. But cherries are such small fruits that they nestle themselves among the blades of grass and it's really a thankless chore to remove them from a lawn.

Birds, unfortunately, often share our taste in ripe fruit. Just as with strawberries and blueberries, we often must compete with birds for our fair share of the cherry harvest. With big trees, there isn't much you can do other than to pick the fruit as fast as it ripens. Small cherry trees can be covered by netting to deter birds. Even this isn't a guarantee you'll save your harvest—you must check and double-check that there are no openings (birds are very persistent creatures).

Probably the biggest and most frustrating problem with sweet cherries is cracking of the fruit. (Sour cherries don't crack, probably because they have much less sugar.) A typical scenario goes like this: After you have carefully pruned, tended and watered your tree for weeks, a bumper crop of sweet succulent fruit is just about ready to pick. Then, clouds gather, it rains—and your crop is ruined! This misfortune results because droplets of water landing on the fruit are absorbed and make it swell, splitting the tender skin. Once the cherries have split, they rot and mold very fast. The sweeter the cherries and the warmer the weather when the rain comes, the more cherries on the tree will crack, and the deeper the cracks will be. There is some consolation in all this—certain varieties are more resistant to cracking than others. We tell you which these are in the Guide to Cherry Varieties.

Besides choosing crack-resistant varieties, there are a few other things you can do to keep your crop intact. Commercial growers hire helicopters to fly over their orchards, blowing the moisture off the fruit before it does its dirty work, but the home grower doesn't have this option. You could, however, use the reverse cycle on your vacuum cleaner to blow water off the fruit of a small tree. Just be sure to be gentle so that you don't knock the cherries off. Some varieties are so prone to splitting that this treatment may not help, but it's worth a try. If rain is forecast overnight, you can cover a small tree with plastic to keep it dry. Don't leave the plastic on during the day, however; a plastic tent could easily turn into an oven. Remove it first thing the next morning. Don't try using one of these tents during a daytime shower; the warmer temperature combined with the moist air creates very attractive conditions for fungus activity.

CHERRY FRONTIERS

After reading this chapter, you can probably guess that researchers are directing their efforts toward producing more rootstocks for cherries, especially rootstocks that will dwarf the scion. Dr. Ron Perry, of Michigan State University, may soon be able to release a dwarfing rootstock, at least to the commercial market.

Dr. Perry isn't the only researcher examining rootstocks. Breeders in Europe are hard at work too, and new cherry rootstocks have emerged from West Germany and the renowned East Malling Agricultural Station. These rootstocks are derived from complex crosses among various wild cherry species. Perhaps someday we'll be able to order a tree on good dwarfing rootstock that hails from one of these European research stations.

Guide to Cherry Varieties

SWEET VARIETIES

Bada—Large fruit, red skin with yellow flesh, midseason. Zones 5–8. Cans well.

Bing—Large, nearly black, sweet, midseason. Zones (5)6–7(8). May crack. Very productive. Cans well.

Black Tartarian—Medium, purplish black, very good quality, early. Zones 5–7. Vigorous, hardy, productive.

Cavalier—Medium to large, black, early. Zones (4)5–7. Very resistant to cracking.

Compact Lambert—Extra large, dark red, sweet and juicy, late. Zones 5–7(8). May crack. Vigorous, 8–12 feet tall, hardy, heavy bearer. Cans well.

Compact Stella—Dark red, sweet, very good quality, midseason. Zones (5)6–7(8). Resists cracking. Moderately hardy, productive, self-fertile, genetic dwarf, 10–14 feet tall. Cans well.

Hedelfinger—Medium to large, black, sweet, firm, midseason. Zones 5–7. Resists cracking. Hardy, very productive. Cans well.

Rainier—Extra large, yellow with pink blush, firm, good quality, midseason. Zones 5–7. Resists cracking. Hardy and productive. Cans well.

Republican (Black Oregon, Black Republican)—Medium, black, late. Zones 5–8. Cans well.

Royal Ann (Napoleon)—Large, yellow with red blush, sweet, midseason. Zones 5–7. Very susceptible to cracking. Productive but bud-tender. Cans well.

Sam—Medium to large, black, early. Zones 5–8. Resists cracking. Cans well.

Starkrimson—Dark red, midseason. Zones 5–8. Self-fertile, genetic dwarf, 10–14 feet tall. Most widely adapted compact variety.

Van—Medium to large, dark red, very good quality, late. Zones 5–8. Resists cracking. Vigorous, productive, less bothered by birds.

Viva—Medium, dark red, early. Zones 5–7. Resists cracking. Cans well.

SOUR (PIE) VARIETIES

Meteor—Large, light red, late. Zones 4–7. Very hardy, good for North and high altitudes, 12–14 feet tall. Cans well.

Montmorency—Large, bright red, fine pie cherry, tart and moderately acid, late. Zones 5–7. Resists cracking. Vigorous, hardy, productive. Cans well.

North Star—Medium, wine red, similar to Montmorency but smaller fruit, early. Zones 4–8(9). Resists cracking. Genetic dwarf, 6–9 feet tall. Cans well.

Suda—Dark red, late. Zones 4–7. Good for rainy areas, since resistant to leaf spot. Cans well.

CHAPTER 10

PLUMS AND PRUNES
A RAINBOW OF COLORS

Vital Statistics

FAMILY
Rosaceae

SPECIES
Prunus domestica (European)
P. salicina (Japanese)

POLLINATION NEEDS
European plum is self pollinating.
Japanese plum requires cross-pollination.

HOW LONG TO BEARING
European plum   4–5 years
Japanese plum   3 years

CLIMATE RANGE
European plum   Zones 5–7
(marginal in Zones 4 and 8)
Japanese plum   Zones 5–9 with some
varieties in Zone 4

If you got excited as a little kid by the array of colors in a box of crayons, you'll love growing plums. No other garden fruit comes in such a variety of hues. You have your choice of plums with golden, blue, red, purple or green skins and yellow, red or green flesh. If your impression of plum colors and flavors is based on what you find in the supermarket produce department, you're in for a wonderful surprise when you grow your own. Because plums are so very perishable and delicate when ripe, they're shipped to produce markets while still on the underripe side. This makes buying a flavorful plum from a store almost impossible. Once you've savored a homegrown, tree-ripened plum, you'll swear that what they sell in stores isn't worthy of the same name.

People often wonder what the difference is between plums and prunes. The fact is that prunes are merely one kind of plum, with purple skin, relatively dry flesh and sometimes a freestone pit. Prunes lend themselves to being dried whole because, when ripe, they contain enough sugar to inhibit the activity of microorganisms that produce fermentation. The juicier, less sweet plums don't have as much sugar, so if you want to dry them you must pit them first. If you leave the pits in they'll ferment in the center.

WHICH PLUM IS WHICH?

Nursery catalogs are often confusing in their descriptions of plums and no wonder—there are several different types. The most common is the European plum, *Prunus domestica*, which originated in western Asia, near the border with Europe. Horti-culturists believe that the European plum is a hybrid between two other species—*P. cerasifera,* which has red skin and yellow flesh, and *P. spinosa,* with blue skin and green flesh. It seems entirely possible that the varied fruit colors of European plums could be due to their multihued hybrid ancestry. European plums have been with us for a long time—over 2,000 years. This type forms the largest tree, and most of the plums available in the supermarket are European varieties. All prunes are varieties of the European plum, but not all European plums are prunes.

Although many nurseries list it as a European plum, the Damson plum is actually a different species. It belongs to *Prunus insititia,* which is native to eastern Europe and western Asia. The rather small trees produce small but tasty purple fruit. Damson plums make great jams and preserves, but they're generally not eaten fresh because they're far too tart. The exception is the Mirabelle plum, a yellow variety of *P. insititia* which is sweet enough to eat right from the tree.

The other major group of plums belongs to the species *Prunus salicina,* commonly called the Japanese plum. This is something of a misnomer, for the tree actually originated in China. However, the Japanese cultivated it and developed many varieties. In the United States and Canada today, most varieties sold as Japanese plums are actually hybrids developed by Luther Burbank between this species and various wild Asian and/or wild American plums. These hybrid varieties produce a delicate-looking tree with pointed leaves free of the soft hairs which cover the undersides of European plum leaves. Japanese plums can have red, green, yellow, purple or black skin. The one color you'll never find is blue. Inside, the flesh may be yellow, red or amber.

The American plum, *Prunus americana,* which Burbank used in many of his crosses, is native to the eastern two-thirds of the United States. Its most noteworthy characteristic is that it's much hardier than European or Japanese varieties. This wild plum isn't usually cultivated, and the fruit it produces varies from tree to tree.

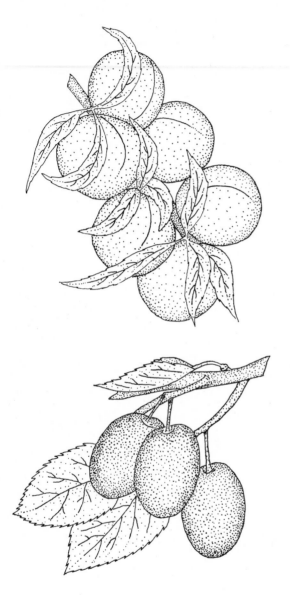

Plum Portrait: The easiest way to tell a Japanese plum tree from a European plum tree is to look at the leaves. Japanese plum leaves, shown at the top, are elongated and rather pointed. European plum leaves are rounder with velvety hairs on the undersides. Plums are available in a rainbow of colors, ranging from red, green, yellow and purple to blue.

The beach plum, *P. maritima,* is a wild plum found only on the East Coast, and is beloved by gourmet jelly and jam enthusiasts for its tart fruit. There are several other wild American plums as well, which are significant because they've been used to breed hardiness into Japanese plums.

For simplicity's sake, in this chapter when we used the word plum, we're referring to both European and Japanese plums, unless we state otherwise.

PLUM VARIETIES

Now that we've sorted out the differences among the various types of plums, we'll give you some guidelines on how to decide which type is for you and which specific variety might suit your needs. (Be sure to turn to the Guide to Plum Varieties at the end of the chapter for more information.)

Climate is an important factor to consider when you're deciding which plum type to grow. In most parts of the country, both European and Japanese plums will grow well. But in the North, Japanese plums tend to emerge from their rest period too early and are vulnerable to late winter damage. Even if they do make it through the winter, they're likely to bloom too soon to escape the clutches of late spring frosts. Japanese varieties jump the gun because they require only 500 to 900 chilling hours. Northern growers would be much better off with European types or the hybrids developed between Japanese and American plums, since they need 700 to 1,000 hours. A good rule of thumb is that Japanese plums will be successful in areas where peaches can be easily grown.

In harsh northern climates, *Prunus americana* hybrids may thrive where even European plums are reluctant to grow. La Crescent, Underwood, Pipestone, Fiebing, Redglow, Superior, Toka, Elliot, Ember and Assiniboine are hardy varieties, but they can be a little difficult to track down, since they're not widely available.

Plums can be grown in the Deep South, but your best bet is to choose from American species rather than from European or Japanese varieties. The Chicksaw plum, *Prunus angustifolia*, which grows from Ohio all the way into the Deep South, has been crossed with Japanese plums to yield the varieties Terrill and McRae. Downing is a variety selected from the wild goose plum, while Wayland was developed from another southern wild plum, *P. hortulana.*

Once you've matched the right plum type to your climate, it's time to think about pollination. Plum pollination, just like cherry pollination, becomes a bit complex due to incompatability problems. There are three basic rules to keep in mind. European and Japanese plums do not cross-pollinate; most European plums are self-fruitful enough to pollinate 30 percent of their flowers, which translates into enough fruit for a bountiful crop; and most Japanese and hybrid plums require cross-pollination.

Redheart, Santa Rosa and Wickson are good pollenizers for most other Japanese plums, while Red Beaut, Kelsey and Eldorado do a poor job. When you grow one of the latter three varieties, you must count on having at least two other compatible varieties so that there's enough pollen to go around. To add another twist, Kelsey and Eldorado won't pollinate each other. Santa Rosa, Beauty and Methley can set a small amount of fruit by themselves (probably just enough for fresh eating) but will produce much better when cross-pollinated.

Commercial growers often solve pollination problems with Japanese plums by grafting a branch of the plum variety Myrobalan 5Q to each tree. Because the Myrobalan 5Q is an abundant pollen producer, each tree ends up having its own effective pollenizer right on the same trunk. If you're not willing to do some grafting of

Matchmaking
Finding Pollinators
for Japanese and Hybrid Plums

VARIETY	WILL NOT POLLINATE	WILL PARTIALLY POLLINATE	WILL POLLINATE
Abundance	Itself		All other plums in list
Beauty		Iself, Santa Rosa	All other plums in list, especially Eldorado
Burbank	Itself, Shiro		All other plums in list, especially Abundance, Beauty, Santa Rosa
Eldorado	Itself	Santa Rosa	All other plums in list, especially Beauty
Elephant Heart	Itself	Beauty	All other plums in list, especially Redheart, Santa Rosa
Frontier	Itself or any other variety		
Methley		Itself	All plums in list, especially Burbank, Shiro
Ozark Premier			All plums in list, especially itself, Methley, Satsuma
Pipestone	Itself		All other plums in list, especially Superior
Redheart	Itself		All other plums in list, especially Elephant Heart, Ozark Premier, Santa Rosa
Santa Rosa		Itself	All other plums in list, especially Methley, Redheart
Satsuma	Itself		All other plums in list, especially Beauty, Santa Rosa

VARIETY	WILL NOT POLLINATE	WILL PARTIALLY POLLINATE	WILL POLLINATE
Shiro	Itself		All other plums in list, especially Beauty, Ozark Premier, Red-heart, Santa Rosa
Starking Delicious	Itself		All other plums in list, especially Ozark Premier, Redheart
Superior	Itself		All other plums in list, especially Pipestone

your own, you really need to plant two different Japanese varieties that are compatible unless you're lucky enough to have neighbors who are growing just the right pollenizing trees next door.

Most European varieties are self-pollinating. There are a few varieties, such as President and Standard, that do require pollenizers. Stanley and Italian are both good for this purpose. President should not be planted too far south, for it needs more hours of winter chilling than most other European varieties. In some California plantings, President drops its fruit buds and leafs out late following a warm winter.

ROOTSTOCKS USED FOR PLUMS

Plums can be grown on a wide variety of rootstocks. Besides growing on various kinds of plum roots, including those of certain wild species, plums can also be grafted onto peach rootstocks.

PEACH ROOTS

Some growers believe that plums growing on peach roots bear earlier and more consistently than those with plum roots. They also point to the peach's higher natural resistance to bacterial canker and to rodents. But there are some disadvantages, as well. Plum trees on peach roots grow more slowly because the graft union is not as compatible as when a plum scion is joined to a plum rootstock. This slight incompatibility can result in smaller crops. Northern growers should be leery of planting plum trees grafted onto peach roots, for they tend to be less hardy than trees on plum rootstocks.

PLUM ROOTS

Plum rootstocks are more tolerant of wet or heavy soils than just about any other fruit rootstock around. But that doesn't mean they *like* these kinds of conditions. You really can't expect any tree to do well if it's mired in waterlogged soil. So even though we say that plum roots are somewhat tolerant of excess moisture, you should still make the effort to provide them with well-drained soil.

The most common plum rootstocks are Myrobalan seedling, Myrobalan 29C (a clonal rootstock) and Marianna 2624. The Myrobalan rootstocks (often shortened to Myro) come from one of the species believed to be an ancestor of the cultivated European plum. Seedling Myrobalan is hardy and well anchored. Since it's grown from seed, Myrobalan varies considerably in other traits, such as disease resistance. This makes it hard to give you a precise idea of what sorts of characteristics a particular rootstock will have. There is one trait that's common to all seedling Myrobalan rootstocks, but unfortunately it's not a good one—susceptibility to oak root fungus.

All Myrobalan 29C rootstocks are genetically identical, so we can confidently give you a rundown of traits that rootstocks of this type will share. These vigorous roots are resistant to root knot nematodes and mildly resistant to oak root fungus, crown gall and crown rot. Unfortunately, they are susceptible to bacterial canker and verticillium wilt. In addition, gophers seem to relish gnawing on Myrobalan 29C, so be sure to protect your plums when you plant them, as we describe in chapter 6 under Getting Fruit Trees into the Ground. Although these rootstocks are vigorous growers, they do take a while to get going. This means the roots will be very shallow during the first three to four years. Compensate for this by careful watering, fertilizing and staking.

Marianna 2624 is a very popular rootstock which is a hybrid between Myrobalan and the wild goose plum. This rootstock is completely immune to root knot nematodes and is moderately resistant to oak root fungus and crown gall. One drawback is that Marianna 2624 is highly susceptible to bacterial canker.

St. Julian, a seedling rootstock, and St. Julian A., a clonal rootstock, are also used for plums. They are derived from the Damson plum, which is a small, compact tree. This trait is carried over into the scion, creating a semidwarf tree.

If you're a small-space gardener you'll be glad to know that there are dwarf plums available. These dwarfs have usually been grafted onto seedlings of *Prunus besseyi,* the sand cherry. Northern growers will be pleased to learn that these rootstocks are very hardy. Trees on *P. besseyi* roots won't grow much taller than 9 feet, so you can plant them as close as 8 feet apart. They're not well anchored though, so you'll need to stake them. Make sure the stake is strong and tied in more than one spot to the tree, for the scion on these dwarfs tends to overgrow the rootstock, making it top-heavy. Then, if the scion bears too much fruit, it may snap right off, destroying the tree.

While most plum dwarfs have *Prunus besseyi* rootstock, some are grafted onto *P. tomentosa* (sometimes called Brompton). These trees tend to be larger than those on *P. besseyi* roots and need to be planted 15 feet apart.

If you have the space, it's better to plant a semidwarf plum tree than a dwarf one. Dwarf plum trees really don't produce much fruit—usually about half a bushel. If you have your heart set on growing enough fruit for drying or making into fruit leather, that amount won't go very far.

A CLOSER LOOK AT PLUM FLOWERS

A plum tree in bloom is a thing of exquisite beauty. Each branch holds aloft a cloud of white blossoms so lovely that you can almost forget the reason for spring's dazzling display—which is of course late summer's bounty of fruit! If you examine a plum branch closely, you'll see that the flowers (and later the fruit) are produced either on lateral buds on one-year-old wood or on spurs. (Cherry tree flowers follow the same arrangement.) Japanese types are more likely to set fruit from buds on year-old wood, while European varieties tend to fruit almost exclusively on spurs. The spurs on Japanese plum trees are 2 to 4 inches in length—longer than those on European plums.

No matter where they develop, plum flowers blossom from buds containing one or two flower buds and no leaf buds. The flower buds on year-old wood appear only at the three or so nodes closest to the base of the shoot. Leaf buds may form at those same nodes, but they are housed in different buds. The arrangement is different for trees that bear on spurs. When each spur grows in the spring, leaves form at the tip. As the spur continues to grow, the flower buds for the next year form in the axils of the leaves. Plum spurs live from five to eight years.

Japanese plums produce three to five flower buds per node, with each bud containing three flowers. Simple multiplication reveals that one node can develop as many as 15 flowers! It's easy to see why Japanese plums tend to overbear. They are very generous in their fruit set and often need to be thinned. European plums generally don't have this problem, but that doesn't mean they'll *never* overbear. Diane found out the hard way that European plums are entirely capable of producing too large a crop. Last year, her Italian prune tree bore a modest crop of large, sweet prunes. This year, her tree (which has never been pruned) was blanketed with lovely white blossoms. Diane thinned the fruit so that there was only one prune on each spur, but there were so many fruitful branches, even that amount of thinning wasn't enough. Because there was simply too much fruit on the tree, the prunes she harvested were only about two-thirds the size of those she enjoyed last year and they weren't nearly as sweet.

When you look at a plum tree laden with a ripening crop, the thing you're probably most concerned with is the status of this year's harvest. But you should also know that at the very same time, the groundwork for next year's harvest is being laid. European varieties form flower buds from late June to mid-August. Japanese plums begin this process later, around mid-July, but they're finished by early August. There's an overlapping of crops here; both types of plums are developing flower buds for the next year's crop at the same time they're maturing this year's plums. If you allow your trees to overbear as Diane did, you may find that fewer flower buds form for next year's crop because the trees just don't have enough nutrients to support both processes at once. This stress can be enough to push your trees into a biennial bearing cycle.

Like so many other fruits, plums are very susceptible to cold temperatures during flowering and immediately afterward. To show you how touchy they are, here are some results from studies that were done in Oregon plum orchards. Italian prunes would only set fruit well when the temperature was above 60°F after blooming. If the thermometer hovered around 50°F, very little fruit was set. Italian prunes are especially sensitive to cool temperatures because their pollen tubes grow very slowly (about a third as fast as those of some other varieties), even under the best conditions. When it's cool, the tubes slow down even more. Because they're moving at such a snail's pace, they don't manage to reach the egg cell while the female parts of the flower are still receptive—so no fertilization takes place. Fortunately, other prune varieties such as Stanley and Brooks aren't quite as temperature sensitive and will manage to carry out fertilization even when temperatures are on the cool side.

FROM FLOWER TO FRUIT

Once the flowers have been pollinated and fertilization has been successful, plums pass through the same three stages of development other stone fruits do. When more than one plum sets on a spur, the fruits in the cluster will be small since they're all competing for a limited amount of nutrients. The tree itself can correct an imbalance of too many plums and not enough nutrients by dropping some of the excess fruits.

In addition to the usual fruit drop periods, plums undergo a late drop after the fruits have turned color. This is called the blue prune drop. Despite its name, the blue prune drop occurs in both prunes and plums of all colors. It can be especially large if your trees are growing in shallow or heavy soil, which would inhibit good root growth and place the trees under nutritional stress. There are other factors that can increase the blue prune drop, such as high temperatures, water stress or a potassium deficiency. It's pretty obvious that if you make sure your plum trees are well watered and well fed, and provide them with rich soil, you'll be eliminating most of the factors that contribute to an especially heavy drop.

WHEN THINNING IS IN ORDER

After the various stages of fruit drop, you may notice that the plums on a particular tree are closer together than 4 to 6 inches along the branches, or that there's more than one fruit per cluster or spur. These are signs that the tree is overloaded with fruit. You should step in and thin the fruit no later than two months after full bloom (when

50 to 80 percent of the blossoms are open). This way you'll encourage plums of generous size to develop. However, if you can't thin in time to affect fruit size, that doesn't mean you should forget the whole thing. Late thinning is better than no thinning at all. Remember, too many ripening plums will place nutritional demands on the tree that will jeopardize next year's crop and possibly get the tree started on the feast-and-famine cycle of biennial bearing.

In some seasons, following our guidelines for thinning may not be enough. If the plums on your tree are slow to ripen and not as flavorful as those in previous years, you need to thin out even more fruit next year. Instead of leaving one plum per cluster or spur, remove some clusters completely on about one-quarter of the branches on the tree.

PLANTING AND PRUNING PLUM TREES

The amount of space you allow each plum tree depends basically on its growth habit. There's really not a standard tree shape to which all plums conform. It makes sense that plum varieties vary widely in form; after all, they are derived from different species. Burbank, for example, has a spreading shape, while Santa Rosa is more upright. We can make some general observations, though, that will help guide tree spacing. European plums tend to be larger and more upright than the spreading Japanese ones, so space full-size trees of this type 20 to 22 feet apart, depending on the variety. Give Japanese varieties 18 or 22 feet of growing room. You can control the height of both types by pruning. Many growers find that keeping standard trees pruned to 15 feet is convenient. If you allow these trees to go unpruned, they may soar to twice that tall! Semidwarf trees may be planted 15 to 18 feet apart, regardless of which type they are.

INITIAL PRUNING

Plum trees tend naturally to grow in an open center, vase shape (although some spread out a bit more than others, as we mentioned above). When you prune, it makes sense to follow this natural inclination. After the young tree is settled in the ground, head it back to 30 to 36 inches in height to promote lateral branching. Continue with the initial pruning as described under Pruning an Open Center Tree in chapter 6.

THE HANDS-OFF APPROACH TO PRUNING

Once you've established the basic shape of the tree, for the next three or four years your only job is to remove water sprouts. Leave the rest of the tree unpruned. Although you may feel as if you're neglecting your tree, in the long run it's a beneficial sort of neglect. Pruning encourages vegetative growth; for a bumper crop of plums, fruitwood growth is what you want to promote. If you hold back on the pruning, the tree will concentrate its efforts on producing the fruiting wood and will come into bearing one to two years sooner than a pruned tree would.

Once the tree starts to bear fruit, you will probably need to do some careful pruning. Pruning at this time helps to keep fruiting under control. For example, if you live in an area where late spring frosts frequently strike, the tree could fall into alternate-year bearing. When that happens, pruning can help bring the tree back in line. During the dormant period before a season in which you anticipate heavy bearing, thin out some of the secondary branches.

Even if a plum tree is bearing well every year, you will need to prune out some secondary branches by the time the tree is five or six years old. This careful pruning ensures that the tree doesn't bear too heavily, which could put undue strain on the scaffold limbs. When you do this sort of pruning, carefully head back branches to smaller laterals that are growing in a direction that enhances the shape of the tree.

For bigger, earlier plums, prune to encourage more vegetative growth. This makes a tree less likely to overbear. If your tree is laden with lots of plums but they're small in size, that's a sign you should do more pruning the next dormant season. In general, European plums need less pruning than Japanese varieties, since they bear less heavily.

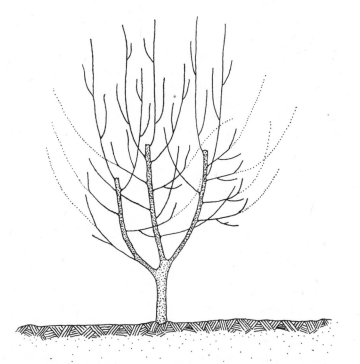

Pruning to Lighten the Load: As your plum tree matures, it may start to bear too abundantly. This can strain the scaffold limbs almost to the breaking point. Keep the tree from overbearing by heading back some secondary branches to small laterals.

WHAT PLUM TREES NEED TO GROW

On the average, European plum shoots should grow anywhere from 9 to 19 inches each year while the tree is young. After the tree has been bearing for two or three years, the shoots should grow about 6 inches annually. Young Japanese plums should grow 10 to 20 inches a year, while older ones usually slow down to around 10 to 12 inches (although some vigorous trees will put on more growth). Using these figures as a basis for comparison, you can gauge whether your trees are growing too slowly. If they are, they may need an extra boost of nitrogen.

The best way to avoid a deficiency problem is to see to it that your trees get a yearly feeding of nitrogen. Starting when the trees are three years old, add 0.05 pounds of nitrogen per tree for every year of age. Stop increasing the dose when the trees turn ten. In chapter 1, under The Nutrients Fruiting Plants Need, you'll find recommendations for good sources of nitrogen, as well as for the other nutrients we discuss in this section.

As we mentioned earlier, a potassium deficiency can lead to an extra-heavy blue prune drop. Another sign that your trees aren't getting enough potassium may at first appear to be a symptom of nitrogen deficiency—small, yellow leaves. But if the leaves go on to form brown, scorched-looking areas at the margins, you'll know it's more than a nitrogen deficiency. Eventually the leaves will turn completely brown. Although it's too late to save these leaves, apply a sprinkling of a good potassium source to ensure that the same thing doesn't happen again next year.

Plum trees are especially sensitive to a zinc deficiency. A tree that's not getting enough will be slow to leaf out and, when the leaves finally do appear, they'll be very small and yellowish. Although they're not as common, other nutrient deficiencies can occasionally pose a problem for plums. Check back in chapter 1 for symptoms of the various deficiencies.

TIPS ON HARVESTING PLUMS
AT THEIR SWEETEST

Capturing plums at their sweetest is a waiting game. You must be patient, watch for signs of ripening, and finally, rely on your taste buds to tell you that your plums are at their sweet and succulent best.

Plums gain considerably in weight and size during the final growth phase, so the later you pick them, the larger they will be. You can begin to estimate your harvest date when the plums begin to develop their final color (this occurs about 20 to 30 days before they're ready to harvest). The sign that harvest is imminent is a slight softening of the fruit. Once your plums start yielding to gentle thumb pressure, they just get sweeter and sweeter. The fruit ripens unevenly on the tree—those that get the most sunlight will ripen first. Take the hint and look for your first ripe fruit among the upper and outer branches. No matter whether you're perched on a ladder or have both feet planted on the ground, use a light touch as you harvest. Pick the plums with a gentle upward twist so you don't damage the spurs.

While you're harvesting, you may come across some plums with cracks at the end of the fruit. During the final rapid growth phase, high temperatures or water stress followed by watering can trigger a burst of rapid growth that literally causes the flesh to outgrow the skin. You can prevent this from happening to a certain degree by keeping your plum trees evenly watered. Sometimes, you may spot cracks along the sides of the fruit, too. These cracks are caused by temperatures that are too cool, so there's really nothing you can do about them except hope for warmer weather.

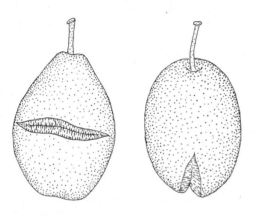

Blame the Weather: Cool temperatures can cause plums to crack along the side, like the fruit on the left. The other extreme, temperatures that are too high, can lead to cracks at the end of the fruit, as shown on the right.

Some seasons you may wonder why your plums don't seem to be as tasty as those you remember from previous years. Before you suspect your taste buds are playing tricks on you, look at your tree. Is it bearing an especially heavy crop? If your tree is overbearing, the fruit won't be as sweet as that from a more moderate-size crop. Because there are so many plums, there aren't enough carbohydrates to sweeten them all. Diane found that out the year her Italian prune tree overbore. She waited as long as possible before picking the fruit, expecting it to become sweeter and sweeter. Periodically, she'd taste one, but she was always disappointed—the plums were hardly sweet at all. As the fruit became softer and softer, it became obvious that it just wasn't going to get any tastier, so she picked it anyway. Even though those plums weren't the best she's ever harvested, the story does have a happy ending—with a little sweetening, they made a tasty fruit leather.

Unlike some fruits such as apples, plums continue to ripen after picking. This is the reason they don't store very well. It's best to use your plums as soon as they come off the tree, but if you can't, it is possible to hold them over for a few weeks. For the best results, give them the ideal storage conditions we describe in chapter 1, under What to Do with Ripe Fruit. Damson plums and wild goose plums store better than Japanese varieties. Don't hold Japanese plums for more than a month or they will lose their wonderful flavor.

SOME SPECIAL POINTERS ON PRUNES

For those of you who are growing prunes, we have some pointers to share that should come in handy when harvesttime approaches. Sometimes it's a bit tricky knowing just when to pick prunes. You want to harvest them while they're at their peak of flavor but not wait until they begin to shrivel on the tree. If you pick them too early or too late and then dry them, they're likely to be discolored or misshapen. The prune industry has funny names for such hapless fruits; the brown prunes that result from premature picking are called chocolates and the swollen prunes that result from late picking are called frogs, bloaters and frog berries. You can avoid this wasted fruit by picking your prunes when they are fully ripe—firm but sweet. When the full-size fruit starts dropping from the tree, that's your cue that it's time to begin picking.

If the weather turns hot right before harvesting, prunes and some plums may develop a brown, soft area near the pit. The cause appears to be rapid metabolism of the tissues brought on by the heat. Because it's buried so deep inside the fruit, the tissue next to the pit can't get rid of the waste products of metabolism. It also can't get enough oxygen, so some of the cells actually die.

This pit burn, as it's called, can also develop once you've picked and stored the fruit. If you store the prunes with the stems removed, pit burn is less likely, probably because there is a route for oxygen to reach the tissues around the pit. Early-maturing prunes such as Richards, Early Italian, Demaris and Milton Early Italian tend to suffer more from this problem than later varieties. These early types also don't store as well as later ones. In fact, their storage life is so short that the fruit will actually begin to break down on the tree before it's been harvested. If you're growing an early prune, heed this advice and watch the fruit especially carefully. It's better to pick a bit early rather than late if you have any doubt about their state of ripeness.

Prunes can also develop pit burn if they're stored between 32° and 44°F, the temperature range of the home refrigerator. With prunes, you really don't have much choice but to eat or preserve them as quickly as possible after harvest. Even if you store them under the best of conditions, as we describe in chapter 1, they won't keep for more than two weeks.

There are two ways to preserve prunes—canning and drying. Drying is the easiest, less messy and less time-consuming of the two. Just leave the prunes in the sun for five to ten days until dry; to dry them more quickly, use a food dryer set at 165°F.

PROBLEMS WITH PLUMS

There are several pests and diseases you should guard against. Careful inspection of your plum trees throughout the season will help you stay on top of any problems that may be developing. In this section we've highlighted some of the most common problems you're likely to encounter.

Bacterial Canker: If you spot gum exuding from the crotches of limbs or from branches, you should suspect bacterial canker (also called bacterial gummosis or sour sap). This disease hampers the normal functioning of the tree by plugging up the xylem and phloem. Younger trees have less xylem and phloem than older ones so they're more drastically affected by bacterial canker. Because this disease is caused by bacteria, there isn't much you can do to prevent it. Fortunately, trees can fight this disease on their own and may overcome it. That doesn't mean you should just stand on the sidelines—there *are* things you can do to help your tree along. You need to baby an infected tree, providing the nutrients it needs and keeping it well watered.

Brown Rot: This fungal disease infects blossoms under wet or humid conditions, and can be spread by the plum curculio beetle. There's an easy-to-spot sign that a tree has been infected—the blossoms turn black. In severe cases, twigs are girdled by a toxin produced by the fungus. This girdling cuts off the xylem and phloem, eventually killing the twigs. Gum may also form on infected twigs. If you're growing a blue plum variety and notice that the fruit has shriveled and brown spots have appeared on the skin, recognize these as signs of a brown rot infection.

In order to be able to control this disease once it shows up, you need to know how it perpetuates itself. The fungus overwinters in affected fruit that is left hanging on the tree. (These shriveled plums are descriptively known as mummies.) Blighted blossoms that remain on the tree can also carry the fungus through the winter. If you scrupulously remove all these potential hiding places and carefully dispose of them, there's less chance the fungus will be around next season to bother your trees. Besides removing all affected blossoms and twigs and any mummies you see, clear away any fallen fruit from under the ailing tree.

Flathead Borers: A plum tree that exudes gum from the base is being attacked by flathead borers, which drill their way into the tree. If the tree is badly infested, you really have no choice but to remove it and destroy it. These borers generally don't attack a healthy, vigorously growing tree, so try to keep your plums in the best shape possible.

Plum Curculio Beetle: The most common pest that besets plum trees is the plum curculio beetle. This small, rather homely insect is brown and has a prominent snout. When the fruit forms, the beetle feasts on the flesh and lays its eggs inside. After the eggs hatch, the beetle larvae in turn feed on the plum flesh. Infested fruit falls from the tree prematurely, and the larvae burrow into the ground and pupate. In about a

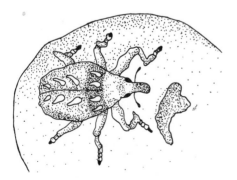

Plum Predator: The plum curculio beetle is the most common pest you're likely to encounter on your plum trees. Keep the ground clear of fallen fruit and trap adult beetles to control this insect.

month, they emerge as adults. These adults feed on the fruit and then depart until the next spring.

If plum curculios have been helping themselves to your plums, one important control measure is to remove all the dropped fruit each day. In addition, you can lay a tarp on the ground under the tree and rap the branches sharply, causing the adult beetles to drop onto the tarp where they're easy to dispose of. While both these measures may control the plum curculio population somewhat, if you're facing a heavy infestation, they may not be enough. In that case you could try using pyrethrum, which some growers find effective against this pest.

PLUM FRONTIERS

New varieties of plums are being released regularly, so you should keep checking what's available in the latest nursery catalogs. Currently, breeders have been putting emphasis on hardier plums and those that will grow well in the South. Rootstocks are another area under investigation. At the University of California at Davis, an intensive study of *Prunus besseyi* rootstock clones is under way to find out which clones are best for grafting onto which plum varieties, so that the problem of the scion outgrowing the rootstock can be overcome.

Guide to Plum Varieties

EUROPEAN VARIETIES

Bluefre—Extra large, blue, freestone, late. Zones 5–7. Bears heavily, vigorous. Use fresh, can, dry.

Damson—Small, blue-purple, late. Zones 5–7. Small tree, different species from other plums. Can, cook.

Early Italian—Large, purple, early. Zones 5–7(8). Use fresh, can, dry.

Green Gage (Reine Claude)—Medium, greenish-yellow, fragrant, mid-season to late. Zones 5–8. Medium-size tree, productive, hardy. Use fresh, can, cook.

Italian Prune (Fellenberg)—Large, purple, freestone, very sweet, late. Zones 5–7(8). Medium vigor, hardy. Use fresh, can, dry.

President—Large, blue, late. Zones 5–7. Needs pollinator, such as Stanley. Cook.

Stanley—Medium to large, blue-purple, freestone, sweet, late. Zones (4)5–7. Vigorous, spreading tree. Use fresh, can, cook.

Sugar—Large, red-purple, very sweet, midseason. Zones 5–7. Medium vigor, tends to be alternate-year bearer. Use fresh, can, dry.

Yellow Egg—Medium to large, yellow, juicy, late. Zones 5–7. Productive, hardy.

(continued on next page)

Guide to Plum Varieties
Continued

JAPANESE AND HYBRID VARIETIES

Abundance—Yellow with purple blush, early. Zones 5–9. Good pollinator.

Beauty—Medium to large, red, midseason. Zones 5–9. Use fresh.

Burbank—Large, red-purple, meaty and firm, midseason. Zones 5–9. Vigorous, low-growing, drooping tree. Can, cook.

Eldorado—Large, dark red, early. Zones 5–9. Upright tree. Can.

Elephant Heart—Extra large, purple, juicy, late. Zones 5–9. Hardy, productive, upright. Use fresh, can cook.

Frontier—Large, blue-black, excellent flavor, midseason. Zones (4)5–9. Vigorous, productive.

Methley—Small to medium, red-purple, juicy, sweet, early. Zones 4–9. Good pollinator, low chill, upright.

Ozark Premier—Extra large, red, midseason to late. Zones 5–9. Hardy, productive. Use fresh, can, cook.

Pipestone—Large, dark red, early. Zones (4)5–9. Has sterile pollen. Use fresh, cook.

Redheart—Medium, red, midseason. Zones 5–9. Use fresh, can, cook.

Santa Rosa—Large, purple, fragrant, early. Zones 5–9. Large, vigorous, upright. Most popular variety. Use fresh, can, cook.

Satsuma—Large, dark red, sweet, midseason. Zones 5–9. Moderately hardy, upright. Use fresh, can.

Shiro—Medium, yellow, early. Zones 5–9. Vigorous, low-growing. Use fresh, can, cook.

Starking Delicious—Medium to large, dark red, very juicy, midseason. Zones 5–9. Vigorous, hardy, disease-resistant. Use fresh, can.

Superior—Extra large, dark red, late. Zones (4)5–9. Hardy, grows quickly. Use fresh.

CHAPTER 11

PEACHES AND NECTARINES
BURSTING WITH FLAVOR

Vital Statistics

FAMILY	**HOW LONG TO BEARING**
Rosaceae	1–3 years
SPECIES	**CLIMATE RANGE**
Prunus persica	Zones 5–9 with some varieties in Zone 10 (marginal in Zone 4)
POLLINATION NEEDS	
Most varieties are self-pollinating.	

It's hard to imagine anyone *not* wanting a peach tree or two in his or her yard. They're among the most ornamental of all fruit trees, with their clusters of long, slender, shiny deep green leaves and their lovely spring blossoms. The fruit itself is ambrosial—just remember the last time you sank your teeth into the luscious, silken flesh of a juicy-ripe peach. Unless you're growing your own, however, you're missing out on knowing just how good a peach can be. Because ripe peaches bruise so easily, it's very difficult to find good, tree-ripened peaches in the supermarket in most parts of the country. Good nectarines are even harder to find. Even gardeners with only a limited amount of space at their disposal can know the joys of homegrown peaches and nectarines; these trees come in a variety of sizes, all the way down to tiny genetic dwarfs that can be grown in patio tubs.

FROM ASIA TO NORTH AMERICA— WELL-TRAVELED TREES

Asia is the birthplace for many stone fruits, and the peach is no exception. Today, peaches still grow wild in the central provinces of China. Some of these peaches may seem exotic to those of us who are used to the round, yellow-fleshed varieties. Besides the color we're familiar with, Chinese varieties come with white and even red flesh. They also grow in two different shapes—round or a somewhat flattened oval.

From China, peaches traveled to Europe, spread from one end of the continent to the other by the Greeks and especially by the Romans. The ease with which peaches can be propagated contributed no doubt to their successful leapfrogging from continent to continent. A peach tree grown from seed will almost always produce edible fruit, a valuable asset seized upon by early horticulturists.

In England and France, the initial varieties were mostly white and soft-fleshed, while the varieties developed in Spain were yellow-fleshed and firm. North America had no peach trees until the Spanish conquistadors brought peach seeds to Mexico and Florida. Once there, the trees escaped from cultivation and began to grow in the wild. Native Americans in Mexico and the South also cultivated these peaches, aiding in their spread. The rootstocks dubbed "native" are derived from the descendants of this group of southern wild peaches.

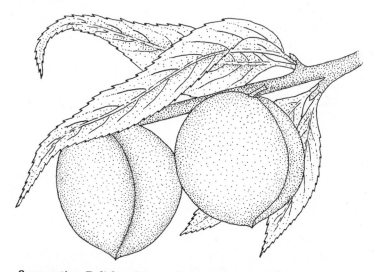

Summertime Delights: Sweet, juicy peaches are certainly one of summer's most welcome treats. They'll be even sweeter and juicier when you grow your own and let them ripen to perfection on the tree.

You may be wondering where nectarines fit into all this talk about peach history. Nectarines appear to have originated from peaches in the Turkestan area of western Asia. If you compare their latin names, you'll find that nectarines are the same species as peaches, *Prunus persica.* This close kinship isn't really surprising, for nectarines are actually just a mutated form of peach. The mutation that leads a tree to bear nectarines rather than peaches is quite common and can occur in the tree as well as in the seed. Sometimes peach trees produce a branch that bears nectarines, but the converse, nectarine trees that revert back to peaches, hasn't been known to happen. This leads scientists to believe that the characteristics of nectarines derive from a genetic flaw. If so, it is a flaw we should be grateful for. Nectarines are fuzzless and have what might be described as a mild but slightly tangy peach flavor. They're also more fragile than peaches.

While it's easy to tell the fruit apart, it's not so easy to distinguish the trees— peach and nectarine trees look exactly alike. A friend found this out when she ordered a nectarine tree from a mail-order nursery. She tended the tree carefully for four years without seeing a single bloom or piece of fruit. Then, when she sold her house, her "nectarine" tree burst into flower, producing a beautiful crop of *peaches* right before she moved. An irate phone call to the nursery resulted in the shipment of a lovely nectarine tree which she planted in her new yard. The fact that peaches and nectarines are so much alike can cause some confusion, as our friend found out. But their similarities are helpful when it comes to talking about how they grow and how to care for them. From now on in this chapter, when we say peach, we'll mean both peaches and nectarines, unless we specify otherwise.

PEACH AND NECTARINE VARIETIES

The most important part of selecting a peach or nectarine variety is to match it to your particular climate. The buds on both these trees are especially sensitive to cold temperatures but the wood is hardier (more about this later, under A Closer Look at Flowering). In general, peaches are hardier than nectarines, both in the buds and in the wood. A glance through the Miller Nurseries catalog, which specializes in hardy fruit varieties, reinforces this idea that peaches can take cold weather better. The 1984 catalog offers 11 peach varieties, but only 3 kinds of nectarines!

Northern growers shouldn't have a hard time finding peach varieties that can withstand the cold. Both the buds and wood of the variety Reliance can survive temperatures of $-25°F$. (Many varieties will die if temperatures fall below $-20°F$.) Another good variety for the north is Madison. Redhaven, one of the most widely planted peach varieties in the world, also does well in the North, although it's not supposed to be as hardy as Reliance or Madison. However, Dorothy has a Redhaven tree which produced after a miserable Montana spring.

For southern growers, the chilling requirement of a variety is more of an issue than its hardiness. However, it isn't always easy to say exactly what the chilling needs are. For a long time, it was thought that peach trees needed 600 to 900 chilling hours below 45°F to break dormancy. This turns out not be true in all cases. In some areas of the Northeast, peaches begin to accumulate chilling hours below 52°F, and in Florida, some varieties start to "count" hours at 68°F! Another complicating factor is that the number of hours may vary from area to area for the same variety. In Fresno, California, July Elberta can get by with only 440 hours of chilling, while in Georgia, the same variety requires 500 hours. The variety Elberta showed an even greater discrepancy, needing only 480 hours in Fresno and 700 hours in Georgia.

Keeping all this in mind, we can only suggest that with peaches more than with any other fruit, it's important to check with your local extension agent to find out which varieties succeed in your area—especially if you have warm winters. A friend of ours in Los Angeles learned this lesson the hard way. She bought a peach tree at a local nursery, only to find that it needed more chilling hours than the balmy southern California climate provided. Because it didn't get enough chilling, the tree was thrown off its usual timetable and was just beginning to leaf out weakly in August. A tree like this won't be able to manufacture enough carbohydrates before the next cool season and will eventually die. Fortunately, growers in mild-winter regions don't have to go through what our friend did, since there is a wide variety of low-chill peach trees to choose from.

Size is another factor you'll probably want to consider as you select a tree. Many nurseries sell grafted semidwarf or dwarf varieties of peaches and nectarines, so space-saving trees are not hard to find. If you're interested in one of these trees, make sure the rootstock it's grafted onto is appropriate for your area (see the following section on peach rootstocks).

Both peaches and nectarines also come in a number of genetic dwarf varieties which grow only 5 to 8 feet tall and usually spread about 6 feet. You'll find that these trees are wonderful landscape accents, for they have such lush leaves, and their small size allows them to fit in just about anywhere. You can also plant them in large tubs and use them to dress up a sunny patio or balcony. If you live in a northerly area where a peach or nectarine tree might not make it through the winter in good shape out in the yard, you can plant a genetic dwarf in a movable tub and bring it into an unheated porch for the winter. There are at least nine genetic dwarf peach varieties and seven nectarines available, several of which are featured in the Guide to Peach and Nectarine Varieties at the end of the chapter.

When it comes to weighing the merits of one peach or nectarine variety against another, there a few other characteristics to consider. Most peach varieties for the home garden are freestone, but a few are semifreestone; with these you have to gently convince the flesh to give up the pit. Only a few cling varieties are available to home growers. If you want to do a lot of canning, clings are the peaches you should grow. (We talk more about clings later under Some Perplexing Peach Questions Answered.) A few nectarines, such as the genetic dwarf Le Grand, are clings. Most peaches and nectarines have yellow flesh, but there are a few white-fleshed varieties, which are every bit as tasty as the yellow ones. Very delicate and mild in flavor, they are considered something of a gourmet item.

Peaches in the Landscape: Genetic dwarf peaches are perfect for container growing. A couple of these trees in redwood tubs make wonderful accents on the patio. Dwarf or semi-dwarf peaches are assets to the landscape as well with their lush clusters of shiny, deep green leaves and sprays of spring blossoms.

PEACH ROOTSTOCKS

With peaches, it seems that each nursery has a favorite rootstock (or rootstocks) for grafting, so which rootstock you end up with is really determined by which nursery you buy from. There's almost a bewildering number of rootstocks that can be used for peaches. One reason for this variety is that no peach rootstock has yet been found that produces a true semidwarf or dwarf tree. For example, Stark Brothers Nursery uses St. Julian plum or *Prunus tomentosa* rootstock for its dwarfs and St. Julian for the semidwarfs. Stark's standard trees are grafted on Elberta seedling, native peach seedling, plum seedling or Myrobalan. Miller Nurseries, however, grafts its dwarfs onto *P. besseyi,* while Burpee uses *P. besseyi* or *P. tomentosa.* (If these *Prunus* species rootstocks as well as the St. Julian and Myrobalan rootstocks sound familiar, that's because they're the same ones used for plums. Refer to the section on Rootstocks Used for Plums in chapter 10 for a review of their characteristics.) Other nurseries use other rootstocks, so the subject can really get quite complicated. To help you sort through this maze of rootstocks, we can only tell you that peaches do best when they're on some sort of peach rootstock rather than on roots of another species. As it is, peaches are rather short-lived trees, and grafting onto a rootstock of a different species only shortens their life-span, since the grafts aren't completely compatible.

Fortunately, some promising new peach rootstocks for standard-size or slightly smaller trees are appearing on the scene. One of these newcomers, Siberian C, was developed at a research station in Harrow, Ontario, and is cold hardy. Not only that, but it can also transmit its hardiness to the scion variety. It does this by inducing the scion to harden early and drop its leaves sooner in the fall. In the spring, blooming is delayed by about five days, which can be a crucial margin in late-frost areas.

Besides this noteworthy hardiness, Siberian C reduces the size of the tree by 10 to 20 percent. For example, Redhaven produces a tree less than 12 feet tall and only 18 feet wide when grafted onto Siberian C. Normally, Redhaven grows at least 15 feet tall and 25 feet wide. Siberian C also brings a tree into bearing at an earlier age than most other rootstocks. Despite all these wonderful attributes, this rootstock isn't perfect. For example, it is susceptible to root lesion nematodes. In heavy soils, Siberian C roots may sucker badly, be poorly anchored and shorten the life of the tree. Southern growers should avoid this rootstock because it frequently doesn't survive the warmer, alternating temperatures of a southern winter. Siberian C also reduces the yield slightly, but this shouldn't really be a drawback; a Redhaven scion on Siberian C still manages to produce an average of 120 pounds of fruit per year. If, after considering the pros and cons, you're interested in growing trees grafted onto Siberian C, see the Plant Source Directory for the listing of Hilltop Orchards and Nurseries.

Here's a word of warning for northern growers. Avoid buying peach trees grafted onto Elberta, one of the less hardy varieties often used as a rootstock. If you live in a more southerly clime, however, you don't need to be as concerned about which rootstock is used for a standard tree. One rootstock which is especially good in the Deep South is Nemaguard, actually a cross between a peach and another *Prunus* species. Nemaguard is completely resistant to nematodes, but it's very susceptible to cold—even some of the winters in North Carolina and Arkansas are too chilly for it.

French breeders are hard at work developing peach rootstocks, and one new one called Amandier looks very promising. A cross between a peach and an almond, Amandier has been quite compatible with all scions of both peaches and nectarines tried so far. This French rootstock is very vigorous and is especially well adapted to heavy, wet soil. It is particularly welcome, for it may completely eliminate replant problems (see the section Problems with Peaches later).

A CLOSER LOOK AT FLOWERING

Peach trees open their lovely pink blossoms in the springtime, later than apricots but before apples. Unlike other stone fruits, peaches produce one flower per bud, and the leaf buds are always separate. You can easily distinguish the flower buds from the leaf buds—the flower buds are slightly swollen at the base, while the leaf buds are elongated and narrow. Most of the flowers you see are on one-year-old wood, with just a few borne on spurs. If you examine a tree in bloom you'll find that most of the flowers are located toward the tips of the shoots rather than near the bases; the basal nodes are more likely to bear only leaf buds. The buds on the very ends open first and go on to produce the largest fruits if they're pollinated. This sequence of blooming means that peaches can be a source of the same sort of frustration as strawberries — if a late frost hits the flowers, those that would have given you the biggest fruits are the ones that are killed.

A tree starts forming flowers for the next season from late June to late July, at the same time it's growing this season's fruit. This means that the tree must support its maturing crop while doing the critical work of initiating flowers. Don't skimp on watering during this time, otherwise you're likely to end up with a poor harvest both this year and next.

Peaches tend to be generous in their flower production — a moderately vigorous tree may initiate over 25,000 flowers! Since there are so many flowers to start with, a peach tree can produce a generous crop even when only 5 percent of those flowers set fruit. This abundance of flowers also helps reduce effects of a late frost, since all the flowers aren't open at the same time. Even if a sudden cold snap hits a tree in full bloom, there will be more flowers coming along to replace those that were caught.

HOW MUCH COLD CAN PEACH BUDS TAKE?

As we mentioned earlier, peach flower buds are quite sensitive to winter's cold, and the buds on nectarines are even slightly more so. If a stretch of warm weather comes along when the buds are ready to break dormancy, any subsequent drop in temperature will do the buds in. The flower buds of peaches are more sensitive to cold than the wood, so peach flowers in the North often are killed by temperatures which leave the wood undamaged. Should a winter with fluctuating temperatures cause you concern about the fate of your peach blossoms, cut through a bud to check on its condition. If it's brown in the center, you can stop thinking about an abundant peach harvest that summer! A few varieties — notably Reliance and Madison — have hardier buds than more familiar varieties such as Elberta. Besides choosing one of these naturally hardier varieties, you can help your peaches grow the most cold-resistant buds possible by nourishing and tending the trees well. In the long run, buds that contain lots of protein and sugar will be hardier than less well-nourished buds.

POLLINATION POINTERS

The good news it that there's not a whole lot to tell you about peach tree pollination, since nearly all varieties are self-fruitful. The exceptions are J. H. Hale, Indian Free and Indian Blood Cling. These varieties produce sterile pollen, so you must plant them with another variety (*not* one of these three) in order to produce crops.

Because peach trees have so many flowers, you don't need to be as concerned with the number of bees buzzing around as you are with apple and pear trees. Only a small percent of the flowers need to be pollinated for you to get an ample harvest. If you have a cloudy, cool spring, pollination may be a problem, though. Dorothy was delighted this spring when two of her peach trees bloomed for the first time. But then a cold snap came, with temperatures in the upper 30s, along with cloudy skies and rain. The bad weather lasted several days, and she ended up harvesting all of three peaches from one tree and none at all from the other.

THINNING MAKES ALL THE DIFFERENCE

While peach trees don't fall into a typical biennial bearing habit, they can develop a pattern of producing more heavily in alternate years. A heavy fruit set will place extra demands on the tree and may jeopardize next year's flower buds. Because its strength has been sapped, the tree will also be very susceptible to winter damage.

The trees themselves make an effort to thin their fruit-laden branches. Peaches, like other stone fruit, have a June drop period when the trees shed some of their excess fruit. But even after that, trees which have had good conditions during blossoming will probably still be carrying too much fruit for their own benefit. If you allow the tree to bear too many peaches, the fruit will be small and bitter tasting because it lacks sugar. One study done in California really drives this point home. Elberta trees which were allowed to carry one fruit for every five leaves produced peaches 2 inches in diameter, with only 5.3 percent sugar in their tissues. Trees with 20 to 30 leaves per fruit bore peaches almost 3 inches in diameter, with 7.6 percent sugar. Elberta trees carrying one fruit for every 50 leaves produced peaches that contained 8.8 percent sugar, while peaches from trees with 75 leaves per fruit had a whopping 9.1 percent!

Clearly, it's in your best interest as well as the tree's to thin the fruit. But before you can do this effectively, you need to understand a little bit about how the fruit is formed so you can time thinning just right.

Peaches come in two basic types—slow ripening and fast ripening. The slow-ripening varieties follow the typical stone fruit pattern of an early growth stage right after blooming, a second stage during which there is little visible change in size, and a final growth spurt in which the cells enlarge and the fruit ripens. Elberta is a typical slow-ripening late variety. The fast-ripening early types, such as Reliance and Redhaven, have a very short second stage. They grow almost continuously, with only about a week of slowed-down growth.

No matter whether the variety you're growing is fast or slow, during the second stage the pit will enlarge the fastest and begin to harden. It's important that you know which type of peach tree you have, for the fruit should be thinned during this second stage, before the final spurt of growth. (See the Guide to Peach and Nectarine Varieties to find out exactly which varieties are fast or slow.) With slow-ripening varieties, you have quite a bit of leeway in choosing a time to thin. But with the fast-ripening types, you must be sure to do the job early enough. A good rule of thumb is to be sure to thin the fruit before it grows larger than 1¼ inches in diameter. There are advantages to thinning when the fruit slows in growth: Most of the natural June drop will have occurred, and the peaches won't cling as tenaciously when you try to pick them.

There are two guidelines you can follow to check whether you've thinned enough. The easiest way is to thin so that the peaches are a certain distance apart.

With early varieties, leave about 10 inches between the fruits. With late varieties, 6 to 8 inches is sufficient. Sometimes this sort of thinning may not be enough, especially in a particularly prolific season. If your tree shows signs of overbearing, you can double-check to see whether more thinning is in order by figuring out the fruit-to-leaf ratio. Aim for about one peach for every 30 to 40 leaves, and you'll strike the proper balance. If counting every leaf on the tree seems a bit overwhelming, choose one part of the tree that has the same ratio of vegetative to fruiting branches as the entire tree. Make an approximation of the number of leaves for each peach.

GETTING PEACH TREES
OFF TO A GOOD START

When you plant your peach tree, be sure to keep in mind all of the criteria given in chapter 1 under Choosing a Location for Your Fruiting Plants. Pay particular attention to avoiding frost pockets—remember how vulnerable to frost peach blossoms are.

If you want your peach tree to do its best, you must provide it with well-drained soil. Poorly draining soil has two bad effects on peaches, as documented in a circular put out by the Ontario Department of Agricultural Foods, called Orchard Soil Management. First of all, it lowers their yield. The yield from trees growing in shallow, poorly drained soil was shown to be anywhere from 30 to 70 percent less than from those growing in well-drained soil. But perhaps even worse, trees stuck in soggy soil have a much shorter life-span than those in well-drained soil. In orchards planted in wet soil, fewer than 19 percent of the trees were alive after 14 years, while orchards with deep, well-drained soils still had 81 percent of the original trees. That's quite a difference!

The usual reason given for steering clear of soggy soil is that the roots suffocate. With peaches, there's another complicating, if not lethal, factor. In waterlogged ground, cells in the peach roots break down and release poisonous hydrogen cyanide, which then kills the roots. A tree suffering from this condition will look wilted—the young shoots will droop, the leaves will turn yellow and drop, and growth will slow down. In severe cases, the tree will die. High temperatures and too much water during the summer can make this problem worse.

If you have any doubts about your soil's drainage, you're better off waiting a year before buying your peach tree. Prepare the soil ahead of time by working in organic matter like compost and coarse builder's sand to a depth of at least 2 feet in an area at least 3 feet square. These soil amendments will increase both fertility and drainage, and ensure that your peach tree has a good foundation to grow from.

You should be careful not to plant your peach tree too deeply. While some fruit trees should be planted somewhat deeper than they were in the nursery, peach trees should be positioned so that the upper roots are only an inch or two below the surface. Peach roots are shallow growers and don't do well if they're forced to grow more deeply—the tree ends up being poorly anchored because the roots don't branch out vigorously. A peach tree, when planted correctly, will have over 90 percent of its root growth in the upper 18 inches of soil. For comparison's sake, a more deeply rooted tree like an apple will have most of its roots in the top 36 inches. Although they're shallower than most, peach roots will spread somewhat farther than those of many trees, to a bit beyond the drip-line.

If you buy a nongenetic dwarf or semidwarf peach tree, be prepared to stake it when you plant it. Since they're not grafted onto peach roots, the grafts aren't very strong. An unstaked tree could split at the graft during a wind storm or under a heavy load of fruit.

How much space you allow between trees depends on what kind you're growing. Standard-size peach trees need 18 to 22 feet of space between them. Trees on Siberian C rootstock, the next size down from standard trees, can be located 16 feet apart. Allow semidwarfs 12 to 15 feet of growing room. If you're raising tiny genetic dwarfs, you can plant them as close as 6 feet from one another or from other bushes or trees in a landscaping scheme.

Because its roots are so shallow, a peach tree forced to compete with weeds or other plants growing in close around the trunk will have problems. If you decide to grow a peach tree in the midst of an expanse of lawn, be sure to follow our recommendations under Planting in Lawns and Other Sites in chapter 6. Keeping the shallow roots evenly moist is critical not only when the tree is newly planted, but throughout the rest of its bearing life as well.

PEACH PRUNING MADE EASY

Here's the ground rule that should guide how you prune and train your tree: Peaches do best when trained to an open center. When you plant the tree, head it back to 30 or 36 inches tall. If there are any branches on the newly planted tree, check out their potential as scaffold branches. Some experts recommend selecting just two scaffolds, while others prefer three or four. Since three or four limbs make for a more attractive tree and give you more branches to work with in the case of winter or storm damage, we think it's better to have more than two scaffolds on a tree. The scaffold branches should be at least 20 inches from the ground and should have an angle of at least 45 degrees. If you're lucky enough to have a young tree with suitable scaffolds, cut these limbs back to 4 to 6 inches to encourage the branching that will give you the fruit-bearing secondary scaffolds. The pruning you do in the tree's second year can proceed as we describe under Pruning an Open Center Tree in chapter 6.

By the third or fourth year, your tree should begin to bear, although it may take somewhat longer in the North. Once the tree starts producing fruit, you must prune carefully. You want to encourage enough vegetative growth so that an adequate amount of wood is produced to bear the following year's crop. But you don't want to encourage *too* much; otherwise the tree will put all its energy into growing leaves and branches. To strike the right balance, follow these general guidelines. Prune out crossing branches and those that are shading others. As you prune, remember to keep

Pruning to Control Production: Keep a tree from overproducing by heading back year-old wood on roughly 20 percent of the branches on the tree. Prune to leave 1 to 2 inches of year-old wood on a branch. If the tree still bears too much fruit, head back more branches the following season.

the center of the tree open. Control the height of the tree by cutting back top branches to outward-growing laterals. As the tree gets older and its growth slows down, you should head back branches that have put on less than 8 inches of new wood a season to outward-growing laterals. And, of course, you should always remove winterkilled or unhealthy branches.

No matter what sort of pruning you're doing, try to prune peach trees as late as possible in the spring—the closer to the time of blooming, the better. Never put the pruning shears to them if there's danger of a cold snap to come that might damage the wood.

If your tree consistently produces too much fruit for the number of leaves, you can correct this somewhat through carefully thought-out pruning. Try heading back about 20 percent of the branches to remove some of the fruiting wood. Wait to see how the tree grows the next season before heading back any more. When you do this pruning, head back so that you leave 1 to 2 inches of year-old wood on each branch. Overbearing is more likely to be a problem for southern growers than for northern ones, because the trees have a longer season in which to put on new growth laden with fruit buds.

SAFEGUARDING PEACH TREES AGAINST DEFICIENCIES

Because their roots are shallow, peach trees tend to be especially sensitive to nutrient deficiencies. In this section, we'll give you the telltale signs of nitrogen and boron deficiencies, since peaches have their own particular set of symptoms. Refer to The Nutrients Fruiting Plants Need in chapter 1 for symptoms of other deficiencies and for information on which organic materials will supply which nutrients.

Since the fruit is borne on the one-year-old wood, peaches need to put on plenty of new growth each year to set the stage for a fruitful harvest the next. The key to lots of new shoot growth is nitrogen. If your tree is seriously deficient in nitrogen, it will have yellow-green leaves and fruit with a bright blush which ripens early. This premature fruit will be watery and lacking in the full rich flavor you'd expect.

The tree's growth rate will also be sluggish. Here are some general guidelines that can help you gauge whether your tree is growing as it should. Established peach trees should have shoots that grow 8 to 18 inches a year, while the shoots of young trees should grow at least 12 inches. Long shoots (24 inches) that have grown too fast generally don't produce many flower buds, nor do shoots on the short side (3 to 4 inches). The shorter shoots also tend to produce smaller fruit.

A very good way to avoid a nitrogen deficiency is to establish a regular feeding schedule. Every spring apply 0.05 pounds of nitrogen for each year of the tree's age. Start when the tree is three years old; once it reaches ten years, stop increasing the dose. Be careful not to overdo it with the fertilizer, though. Too much nitrogen will result in soft fruit with a washed-out color and insipid flavor.

If peach trees aren't getting enough boron, they will let you know. The trees will be late to break dormancy and the fruit will have brown spots scattered throughout the flesh.

HARVESTING PEACHES IN THEIR PRIME

The best way to tell when your peaches are ripe is to rely on two senses: touch and taste. Keep feeling your peaches and when they begin to soften slightly, pick one and taste it. Once the flesh yields slightly to thumb pressure and tastes sweet, your peaches are ready to pick. As with other fruits, the peaches that get the most sun will ripen first, so be sure to check the fruits that are on the upper and outer branches for starters. You'll have to visit your peach tree two or three times, since all the fruit doesn't ripen at once. When you harvest, be gentle, for the ripe fruit bruises easily. Pick peaches with a twist and an upward pull so you don't damage them.

Some fruit growers find it more convenient to harvest all the peaches at one time, once most of them are ripe. Of course, there will always be laggards that are still a little on the underripe side. If a few peaches aren't quite ripe when you pick them, don't put them in the refrigerator. You might be tempted to, thinking that they'll ripen more slowly there and allow you to stretch out your supply of peaches. But you'll be disappointed, for peaches that ripen at cool temperatures just don't taste as good as those that ripen at room temperature. Dorothy had to harvest her ill-fated crop of three peaches before they were completely ripe because the hornets thought they were ready to eat and had begun to sample the fruit. She found that a day or two at room temperature was all it took to bring those firm-ripe peaches into full, juicy sweetness.

Fully ripe peaches spoil quickly, so be ready to use them within a day or two of harvest. They'll continue to ripen after picking, even if you store them in the refrigerator. You really can't count on holding ripe peaches for any considerable length of time.

If you can't deal with a glut of peaches and have suitable cold storage available, there is a way you can carry the harvest over for up to two weeks. Pick some of the fruit when it's almost ripe, meaning yellow but still firm, and store it under the conditions we describe in chapter 1 under What to Do with Ripe Fruit. Check any fruit you store carefully for bruises or any sign of mold. Brown rot and fungi can spread like wildfire through a box of stored peaches and reduce them to mush before you get a chance to enjoy them. Keep in mind that the longer the fruit is stored, the lower its quality will be. Peaches in storage take on a dull color and their flesh may become mealy. We have both bought absolutely awful, cottony peaches from local fruit stands that were most likely left in cold storage too long.

SOME PERPLEXING PEACH QUESTIONS ANSWERED

After you've been growing and harvesting peaches for a few seasons, chances are that you may have observed some curious traits about your fruit—like pits that split and peaches that are lopsided. We'll attempt to shed some light here on a few of the most common quirks you're likely to encounter.

Mushy Home-Canned Peaches: Have you ever wondered why the canned peaches you buy in the store are so firm, while the ones you can at home are soft, almost mushy? The difference is in the peach varieties. Commercial growers who sell peaches to canneries grow cling varieties, which are generally not available to the home grower. Cling peaches are very firm even when they're completely ripe, while the freestone varieties grown chiefly for fresh eating are tender.

To understand this difference, we have to look at the chemistry of the peaches. Freestone varieties have enzymes that dissolve the cement between the cells of the fruit when it ripens. This separates the flesh from the pit and also softens the fruit. Cling peaches lack this enzyme, so the fruit doesn't soften and the pit remains firmly anchored in the flesh. You wouldn't want to use ripe cling peaches for fresh eating because they are so firm, but that same firmness results in a nicely textured canned product.

Most of the peach varieties you'll find in catalogs are freestone. But there are a few cling varieties that are available to the home grower, namely Babygold,

Piedmont Gold and Suncling. (See the Plant Sources Directory for the address of Carlton Nursery.)

Split Pits: Sometimes, the pit of a peach will split when you cut the fruit in half. Split pits are more common in large peaches grown on trees with a light crop than on heavily bearing trees. Other conditions that favor split pits are a big dose of nitrogen and very warm weather during the first growth period, right after bloom. While you may find split pits annoying, you'll also find that the fruits which exhibit the problem are likely to be especially sweet, since they grew under very favorable conditions.

Lopsided Peaches: Occasionally you may find rather comical peaches on the tree with one side fatter than the other. Stone fruit flowers have two ovules. Before a flower opens, one ovule degenerates, and the other is left, ready to be fertilized when the flower opens. The side of the fruit that ends up with the fertilized ovule will often be larger, more highly colored, and even more flavorful than the other side. In some cases it may even ripen earlier. If you cut into one of these lopsided peaches and the pit splits, chances are that the half of the pit bearing the seed will be on the larger half. The uneven distribution of hormones produced by the developing seed are responsible for all of these one-sided traits.

Variations in Color: You've probably noticed as you've been paging through nursery catalogs that some peaches are more colorful than others. Some varieties, such as Redhaven, produce fruit that looks as if it's been bathed in red, while others, like Elberta, have only a slight red blush. Although the coloring of the fruit is predetermined by variety, exposure to sunlight is an important factor. The more sun it basks in, the more highly colored the fruit will be.

You may have observed a reddish area around the pit of certain peaches and nectarines. This is a trait peculiar to a few varieties, and is nothing more than a deposit of red pigments, called anthocyanins. If you can peaches with this reddish area, you'll notice that in about a month the color will leach out into the liquid in the canning jar.

Browning Flesh: You've probably found when slicing up fruit for a salad that the flesh of peaches tends to turn brown. This discoloration occurs when oxygen in the air combines with chemicals in the fruit. Some peach varieties, such as Redhaven, Reliance and the nectarine varieties Flavorcrest, Flamekist, Harko and Nectared 5, are not as quick to brown. The Elberta peach is one of the fastest browning varieties, and Sunglow and Fantasia nectarines are also quick to discolor. The varieties that brown easily also show bruises with a more obvious and distasteful brown color.

PROBLEMS WITH PEACHES

Among the pests to watch out for are the plum curculio and the codling moth. (For more details on the plum curculio, see chapter 10, and on the codling moth, see chapter 7.) Peaches on infested trees will be deformed from the cuts made by the insects and they'll usually fall off the tree before they mature. About the best and only means of preventing an ongoing infestation is always cleaning up any fallen fruit immediately.

Peach trees are susceptible to various fungal diseases, and without the proper care, the chances increase that they'll eventually succumb. For instance, if you don't prune properly, excess foliage may shade too much of the trees and prevent air from circulating freely. This factor in turn may encourage fungal attack, especially in warm, humid climates. For the same reason, don't be tempted to skimp on space and try to squeeze more trees into your yard than there's really room for. Air can't circulate well through a clump of overcrowded trees, and in a humid climate, that's an open invitation for trouble.

What we've given you so far are a few general pointers on problems you should be on the alert for. To help you safeguard the health of your peach trees even more effectively, here are some specifics.

Brown Rot: Like plums, peaches can be infected with brown rot. Since the disease is spread by insects that break the skin to feed on the flesh, keeping down the population of insect pests can help stave off brown rot. If you have a tree with a mild case and only some of the fruit is affected, you can salvage the rest of the crop by harvesting it, even though it's not quite ripe. Leaving the uninfected fruit on the tree to ripen increases the chances that the spores will spread to it and spoil the harvest.

Kill any spores that may be present on the surface of the fruit you pick from an infected tree by dipping the peaches in lukewarm water (submerge them for seven minutes in 120°F water, three minutes in 130°F water or two minutes in 140°F water). This bath will kill the fungi but won't harm the fruit. You can then ripen or store the fruit as described earlier under Harvesting Peaches in Their Prime.

Gophers: These gnawing rodents aren't as fond of peach roots as they are of some other fruit tree roots. Remember, however, that not all peach trees are planted on peach roots. The only tree Dorothy ever lost to a gopher was a semidwarf peach grafted onto a non-peach rootstock. The beast gnawed away every last shred of root, leaving only a stick of a trunk behind. To be safe, it's always best to protect a fruit tree from gophers as described in chapter 6 under Getting Fruit Trees into the Ground.

Peach Leaf Curl: This is one of the worst fungal diseases a peach tree can get. When a tree comes down with this disease, the young leaves turn red or purple, pucker and as the name implies, arch or curl as they continue to grow. Later, they take on a yellowish cast with a silvery sheen and fall off the tree. This early loss of leaves prevents the tree from accumulating enough carbohydrates to see it through the winter, leaving it more susceptible to winter injury. Diane, unfortunately, can give a testimonial on just how severe this disease can be. One of her trees died after just two years because it was so weakened by the fungus. A Reliance peach, the hardiest variety of all, certainly could have withstood the worst Missoula winters if only the fungus hadn't sapped its strength.

Once you've diagnosed the disease, about all you can do is to pick off the leaves as soon as they curl so that they don't fall and make your mulch a haven for

Fungus Has Struck: If the peach leaves on your tree start to pucker, arch or curl, suspect the disease peach leaf curl. This is one of the worst diseases that can afflict peach trees.

new spores. Be sure to wash your hands very well after touching infected leaves if you're going to be checking any other peach trees. You don't want to be the one to spread this disease through your own backyard orchard!

Root Lesion Nematodes: These tiny, soil-dwelling worms can cause problems in peach trees such as malformed flowers, leaves, stems and roots. Another sign that nematodes are at work is very slow growth. Some rootstocks are more susceptible than others to nematodes (see the section Peach Rootstocks, earlier), so if these pests are a problem in your area, choose your rootstock accordingly.

Besides carefully selecting rootstocks, you can try another strategy to foil nematodes before you plant. Studies done at the University of Georgia have shown that French marigolds keep soils free of nematodes. To be effective, you should raise the marigolds on the planting site the season before setting the tree in the ground. Chemicals given off by the marigold roots will kill off nematode larvae in the soil, transforming it into a hospitable site for the peach roots.

Short Life-Span: It's sad but true—peach trees tend to die young. Sometimes, a tree that was healthy in the fall will suddenly wither and die back to the roots soon after leafing out. All you'll be left with is a tree skeleton in your yard. The following year the roots may sprout new shoots, but don't be fooled by them. Your tree has perished and you're better off digging it up, roots and all, and replanting in a different spot. This syndrome, referred to as peach tree short life, is caused by a combination of factors. Damage to the bark, usually caused by freezing or by bacterial canker disease results in infection by canker fungi which kill the tree.

As we noted earlier, under Getting Peach Trees Off to a Good Start, trees grown in waterlogged soil live much shorter lives than those in well-drained soil. Even though providing good drainage is something you can control, you're still more likely to have to replace a peach tree in your lifetime than any other fruit tree. In the South, peaches live only about 8 years, and they rarely survive longer than 18 years in the North. This difference in life-span is due to the fact that the South has more disease problems because the winters aren't cold enough to kill off the insects and disease bacteria or fungi.

Replanting Problems: The fact that peach trees often have to be replaced makes it especially unfortunate that they sometimes don't do well when planted where another peach tree used to be. While no one is certain of all the reasons for this phenomenon, there are several factors involved. Perhaps hydrogen cyanide released by the roots of the first tree lingers in the soil and harms the second one. Or perhaps nematodes that were living off the first tree infest the second. Certain rootstocks, such as Amandier, seem less susceptible to this problem. The best way to avoid endangering a new tree, of course, is to replant in a totally different spot. In a small yard, this won't always be possible. The next best thing is to plant a tree on a different rootstock, preferable one like Amandier.

PEACH FRONTIERS

While there's a lot of diversity among the peaches of the world, those in North America were basically derived from a very narrow selection of varieties. In 1850, Charles Downing introduced a variety called Chinese Cling from England, which was then used to develop the varieties Bell, Elberta and J. H. Hale. To date, almost all our present varieties are descended from this small band of peaches. This is not the ideal situation, for it doesn't give breeders a great deal of genetic variation to draw from in selecting new varieties. The task breeders face today is trying to find genetic lines from other parts of the world to use in creating new peaches. They've started to tap various other sources, so that much headway is actually being made in developing good new peach varieties.

Breeders understand a great deal about how such characteristics as dwarfness, pollen fertility, flesh color, freestone versus cling traits and fruit shape are inherited. This allows them to develop exciting new peach varieties once the proper genetic material is available. One area they need to know more about is how disease resistance is inherited. When they unravel the mysteries of how these traits are transmitted, we can look forward to trees that help themselves stay healthy.

Research on peach rootstocks is just beginning to gain momentum. This is a very important area, for we've seen how serious the short life-span of peach trees and the other problems brought about by inadequate rootstocks can be. Unfortunately, rootstock research is painfully slow, since the stock must be tested over a period of years to see if it is promising, then cloned, grown and retested under a variety of climatic conditions. All of this testing can take ten or more years to carry out. But don't lose hope—there are already promising new rootstocks like Siberian C and Amandier being used for trees sold to commercial growers. If we all have a little patience, these rootstocks and we hope other new, improved ones will start to show up in our nursery catalogs.

Guide to Peach and Nectarine Varieties

STANDARD PEACH VARIETIES

Belle of Georgia—Large, aromatic, white flesh, freestone, late. Zones 5–8. Vigorous, hardy, very productive. Use fresh, freeze.

Cresthaven—Medium to large, yellow flesh, freestone, late. Zones 6–8. Hardy, productive. Use fresh, freeze, can.

Elberta—Large, yellow flesh, freestone, very late. Zones 5–8. Medium vigor, productive. Use fresh, freeze, can.

Florida King—Extra large, clingstone, early. Zones 8–9. Use fresh.

Golden Jubilee—Medium, oblong, yellow flesh, freestone, strong flavor, midseason. Zones 5–8. Freeze, can.

Harbrite—Medium to large, midseason. Zones 4–7. Use fresh, freeze, can.

Indian (Blood)—Medium, clingstone, late. Zones 5–8. Low chill, needs pollinator. Can.

J. H. Hale—Extra large, golden yellow flesh, delicious, very late. Zones 5–8. Medium vigor, moderately productive, needs pollinator.

Madison—Medium, yellow flesh, freestone, late. Zones (4)5–8. Medium vigor, heavy bearer, tolerant of late frost. Use fresh, freeze, can.

Nectar—Medium to large, white flesh, midseason. Zones 5–8. Good in South.

Redhaven—Medium, yellow flesh, freestone, good tasting, early to midseason. Zones (4)5–8. Spreading, vigorous, productive. Widely planted variety. Use fresh, freeze, can.

Reliance—Medium, round, yellow flesh, freestone, early to midseason. Zones (4)5–8. Very hardy, good for North. Use fresh, freeze, can.

Rio Oso Gem—Large, aromatic, yellow flesh, freestone, high quality, very late. Zones 5–8. Productive, touchy but worth the trouble. Use fresh, freeze.

Stark Gulf Queen—Early, yellow flesh. Zone 9(10). Developed for the Deep South, needs only 150 chilling hours. Use fresh.

Stark Saturn—Flattened shape, white flesh, midseason. Zones 5–8. Use fresh.

Suwanee—Large, red blush, yellow flesh, freestone, excellent flavor, early. Zones 7–9. Low chill, very good for Deep South.

Veteran—Medium, midseason to late. Zones 5–7. Blooms late. Freeze, can.

GENETIC DWARF PEACH VARIETIES

Bonanza—Medium, red blush, yellow flesh, freestone, early. Zones 5–8.

Golden Gem—Large, yellow flesh, freestone, firm with good flavor, early. Zones 5–8. Good for landscaping.

Golden Glory—Extra large, mottled red blush, moderate fuzz, yellow flesh, freestone, soft, late. Zones 5–8. Good in cold climates.

Southern Rose—Large, red blush, yellow flesh, freestone, firm, late. Zones 5–8. Low chill, blooms early.

Stark Sensation—Large, yellow flesh, freestone, firm, early. Zones (4)5–8.

STANDARD NECTARINE VARIETIES

Fantasia—Large, yellow flesh, freestone, firm, good quality, midseason to late. Zones 6–9. Low chill.

Flavortop—Large, red blush, yellow flesh, freestone, midseason. Zones 5–8. Vigorous, productive.

Mericrest—Medium, yellow flesh, freestone, midseason. Zones (4)5–8.

Nectared 4—Large, sweet, yellow flesh, semifreestone, midseason. Zones 5–8. Productive.

Redgold (Red Gold)—Large, aromatic, yellow flesh, freestone, midseason. Zones 6–9. Heavy bearer, disease tolerant, hardy, resistant to late frost. Fruit stores well.

Stark Gulf Pride—Yellow flesh, freestone, early. Zones 9(10). Developed for the Deep South, has low chill requirement. Use fresh, freeze, can.

Sunglo—Medium to large, deep yellow flesh, freestone, midseason. Zones 5–8. Vigorous, hardy.

GENETIC DWARF NECTARINE VARIETIES

Golden Prolific—Large, yellow skin and flesh, freestone, soft, late. Zones 5–8. High chill.

Nectarina—Medium, slight red blush, yellow flesh, freestone, midseason. Zones 6–8.

Stark Honey Glo—Medium, red blush, yellow flesh, freestone, midseason. Zones (4)5–8.

Stark Sweet Melody—Large, red blush, yellow flesh, freestone, outstanding flavor, midseason. Zones (4)5–8. Similar to Redgold.

CHAPTER 12

APRICOTS
VELVETY SOFT AND HONEY SWEET

Vital Statistics

FAMILY	**HOW LONG TO BEARING**
Rosaceae	3–4 years
SPECIES	**CLIMATE RANGE**
Prunus armeniaca	Zones 4–9
POLLINATION NEEDS	
Most varieties are self-pollinating; only a few require cross-pollination.	

The most compelling reason to grow your own apricots is to get the chance to savor these luscious fruits at their very best, bursting with a delicious flavor and so tender they almost melt in your mouth.

Store-bought apricots suffer more than perhaps any other fruit from being picked too early and improperly stored. Hardly anyone nowadays knows how a fresh, tree-ripened apricot tastes. Some sense of the apricot's flavor comes through in jams, dried fruit and juices, but it's easy to wonder how the insipid, wizened spheres you find in the produce section of the supermarket could ever produce such tasty confections. The answer, of course, is that there's no resemblance at all between a fruit that's been picked when truly ripe and one that's green and hard as a rock. Once they're picked, apricots don't increase their sugar content, so the ones you find in produce bins are mere shadows, tastewise, of what they could be. When you grow your own, you can let the apricots ripen on the tree until they develop a wonderful honeylike sweetness.

Besides their wonderful fruit, apricot trees are very beautiful in blossom, with their dark gray bark and pinkish flowers. Once they're done blooming, the glossy dark green leaves and reddish branch tips make decorative trees that are assets to the landscape.

Two Seasons of Beauty: Golden apricots against a backdrop of glossy green leaves just before harvest are every bit as beautiful as the blossoms in spring.

SOME APRICOT BACKGROUND

Like so many other fruits, apricots come originally from China and various parts of western Asia. They were grown by the Romans as early as the first century A.D., so they have a long history of domestication. Cultivated apricots belong to the species *Prunus armeniaca,* but within this one species there are trees with quite varied characteristics. This wide range of traits is due to the fact that apricots originated in mountainous regions. Each group of apricots in its own mountain valley became isolated from other groups. Over years of continuous inbreeding, each group in each valley developed its own particular set of characteristics. This explains why there are orange, yellow and almost white apricots; apricots with bitter pits and those with edible, almondlike pits; trees that require cross-pollination and those that are self-fertile.

Most of the apricot varieties we grow today were developed from what is called the European group of apricot types, which has a more limited range of characteristics than some of the other groups. Fewer trees were used to start this group, so as a whole they have a narrower genetic base than other apricots. This lack of variability can be seen as a hindrance to future breeding efforts. Another drawback to these European trees, at least in the eyes of home growers, is that they have a shorter chilling requirement than some others. This is fine for gardeners in mild-climate areas but difficult for those of us in the North.

Their bright orange color is a good clue that apricots are an above-average source of vitamin A. Carotene, an orange pigment which is converted into vitamin A by the body, accounts for the color. The more carotene, the brighter the orange coloring. There are 2,700 milligrams (mg) of vitamin A in 100 grams of fruit (roughly what you get when you get two or three apricots). The same quantity of apricots provides a fair amount of potassium (281 mg) and some vitamin C (10 mg).

APRICOT VARIETIES

Choosing apricot varieties for your garden can be tricky because of the way the trees react to winter cold and to fluctuating winter temperatures. Some apricots are extremely hardy—varieties in China survive temperatures as low as −40°F. But even the hardiest trees can suffer winter damage when the temperature fluctuates.

The apricots we find in nurseries aren't quite as hardy as the rugged Chinese trees, but they're still hardy enough to grow well in most of North America. What makes apricot growing rather touchy is the fact that these trees have a low chill requirement, so fluctuating winter temperatures can bring the trees in and out of dormancy. This fluctuation may ultimately take its toll and kill your trees. North American apricots require from 400 to 1,000 hours of winter chilling, with most varieties needing 600 to 700 hours. Some varieties, such as Early Golden, Gold Kist, Earligold and Newcastle, have exceptionally low chill requirements, and so are only suited to the Deep South.

SOME FAVORITE VARIETIES

If you live in an area with mild winters in which temperatures don't fluctuate wildly, Royal (sometimes called Blenheim) is an excellent choice. This is the most popular variety among commercial growers because the fruit is good canned, fresh and dried. Royal apricots are yellow, sweet, firm and large, and the tree produces a lot of them.

As good as this variety sounds, don't try to grow Royal if winter temperatures in your area fluctuate. The tree itself will die or the flower buds will fall off—either way, you won't get any apricots!

Moorpark is another popular variety, with good reason. The fruit is very large and ranges from yellow to orange in color. It has a very sweet flavor that some people find reminiscent of plums. Moorpark is hardier than some varieties but bears rather lightly. The fruit on a Moorpark tree ripens at an uneven rate; commercial growers would see this as a negative trait, but it suits the needs of most home growers who don't like the feast-or-famine situation that often arises with fruit.

The third most common variety, Tilton, has ample-size fruit, orange skin and yellow-orange flesh. Tilton is a heavy bearer with a high chilling requirement, so it's better able to withstand fluctuating temperatures. The flavor isn't as good as Moorpark or Royal; many people find the fruit bland when canned and dried. But in areas where winter temperatures never stay consistently low, dedicated apricot lovers may find that Tilton apricots are better than no apricots at all. (For a profile of other varieties, turn to the Guide to Apricot Varieties at the end of the chapter.)

TAKING POLLINATION INTO ACCOUNT

Some apricots need another tree as a pollenizer, but others do just fine by themselves. Be sure to determine whether you'll need to provide a pollinator as you investigate apricot varieties. To give you some guidance, three popular varieties that need pollenizers are Perfection, Chinese and Riland. Perfection is a hardy tree, but it has a low chill requirement so it's not really suited for areas with fluctuating winter temperatures. If you've got the right climate, you'll be rewarded with an abundance of very tender, yellowish fruits. Chinese (also called Mormon) is frost resistant and blooms late, so it's a good choice for areas often hit with late frosts. The fruit is medium-size, sweet and juicy, but some people find the flavor isn't quite as intense as in other apricots. Riland has tasty, light yellow fruit with a deep red blush. Basically any other variety (outside of this trio) will be a good pollenizer. Three varieties known to be especially abundant sources of pollen are Moorpark, Royal and Tilton.

VARIETIES FOR NORTHERN GROWERS

Sungold and Moongold are two dwarf varieties touted as especially good for the North. They are often sold together since they pollinate one another. While nurseries often describe them as having a fine flavor, the informal evaluations of their taste we've seen (conducted by the North American Fruit Explorers) have been at best lukewarm, and the fruits are on the scrawny side. And, as you will see in the following section, these trees are far from being the answer for northern growers.

Scout, which was developed in Canada, and Goldcot, from Michigan, are probably better choices for those of you living in the North. Other varieties that may be good where temperatures fluctuate are Hargrand, Wenatchee (which also does well in a dry climate), Andy's Delight and Red Sweet. Hilltop Orchards and Nurseries sells several hardy varieties developed at the Harrow Research Station in Ontario, Canada, so you might want to see what they have to offer. (See the Plant Sources Directory for their address.)

OTHER TRAITS TO CONSIDER

In recent years, there have been two interesting developments in apricot varieties. The first is apricots with an edible, almondlike seed. Apricots and almonds are very closely related trees with basically similar fruits. The flesh of the almond, however, is shriveled and not edible; it's the seed within the pit of the fruit that we eat. While most apricots have a bitter-tasting seed, a few varieties, such as Stark SweetHeart, have almondlike kernels inside the pit. Since the sweet kernel flavor is dominant, you can grow one of these trees along with regular apricots, and even when cross-pollination takes place the pits in the SweetHeart fruit will still be sweet. It's interesting that with almonds, it's a different story. Because the bitter taste is dominant, almond growers can't grow a bitter variety along with a sweet one.

The other relatively new development is the genetic dwarf. Two of these on the market are Garden Annie and Stark GoldenGlo. These small trees are perfect for container gardening or sprucing up the landscape.

ROOTSTOCKS USED FOR APRICOTS

Your task of selecting the best apricot variety for your garden isn't complete until you know what rootstock the tree is grafted onto. With this fruit, perhaps more than with any other, the rootstock can be the key to success. Unfortunately, we have mostly bad news to tell you about apricot rootstocks. Because this fruit isn't one of the more popular ones either commercially or in home gardens, few nurseries go to the trouble of growing apricot seedlings for rootstocks. They just use peach or plum rootstocks instead. Since these trees are all so closely related, apricot scions will grow on peach or plum roots, but they don't do as well as they would on their own wood.

PEACH ROOTSTOCKS

When an apricot tree is grafted onto peach roots, the rootstock often grows faster than the scion, resulting in a thicker lower trunk. This weakens the graft union, and when a windstorm comes along, a tree that has been growing for a few years may snap right off at the graft. One grower we know can sadly testify to this. Out of 30 apricot trees, he lost 13 in a single storm, after having nurtured them for three years. Then, he lost all but 6 of the rest the following winter.

Beware of Peach Rootstocks: One of the possible consequences of grafting an apricot tree onto peach roots is an
uneven growth rate. Sometimes the rootstock will outgrow
the scion, creating a thickened lower trunk. This weakens
the graft, leaving the tree vulnerable to storm damage.

It's unfortunate, but when nurseries use peach rootstock, they often choose less
hardy varieties such as Elberta. It doesn't matter how hardy the scion is if the roots
can't take winter cold. If you live in an area with chilly winters, you'd be wise to steer
clear of apricots grafted onto peach rootstocks.

Until recently, both of us have been puzzled as to why we have had such bad
luck with apricots. There are many large, thriving apricot trees in our town which
have obviously survived the worst winter weather western Montana can deal out. Yet
neither of us can seem to grow an apricot tree with any degree of success. Diane has lost

one tree each of Tilton, Moorpark, Perfection and Rival, while Dorothy has unsuccessfully tried to grow Sungold and Moongold, and Garden Annie (the last in a carefully chosen protected spot). We suspect that most of the trees we lost were grafted onto peach rootstocks that couldn't take the bitter cold. With certain trees, such as Tilton and Garden Annie, the cause of death may have been a scion that dehardened and then succumbed to a late winter cold snap.

PLUM ROOTSTOCKS

Plum rootstocks are hardier than peach rootstocks, but they come with their own set of problems. For one thing, suckers crop up continually, so you need to get out your shears often. Also, the fruit of apricot trees grown on plum rootstocks doesn't have the same full flavor as the fruit that's grown on apricot roots. The fruit growing on the inside branches is especially inferior. But there is some good news—plum rootstocks are more tolerant of wet soil than apricot or peach roots.

APRICOT ROOTSTOCKS

Apricot rootstocks produce hardier trees with more vigor than do plum or peach rootstocks. Before we lead you to believe that apricot rootstocks are perfect, we should warn you that they are sensitive to soggy soil, just as peach roots are. When the roots are surrounded by water, they release hydrogen cyanide which can then damage them.

By far the best rootstock for apricots is Manchurian Bush Apricot, which is extremely hardy and generally more tolerant of wet soil than most apricots. It is also more drought tolerant. These particular roots are well anchored, probably because they are more fibrous and bushy than ordinary apricot roots. Manchurian Bush Apricot is also resistant to canker, borers and gummosis. In terms of size, it dwarfs the scion slightly. Trees of this variety are small, inexpensive and purportedly able to withstand winter temperatures down to −50°F, lower than any potential apricot growing region should ever face. Dorothy has had two growing in her yard for five years, and they haven't suffered any winterkill.

WHICH NURSERIES SUPPLY WHICH ROOTSTOCKS

Most nurseries will tell you what rootstock their trees are growing on if you ask. We did a little inquiring on our own and found out that Stark Brothers Nursery uses cold-susceptible Elberta peach for its standard trees; St. Julian A plum, which is moderately hardy, for its semidwarf trees; and Nemaguard plum, which is not very hardy, for its dwarf apricots. Miller Nurseries uses plum rootstocks for all its apricot trees. However, Hilltop Orchards and Nurseries uses Manchurian Bush Apricot exclusively as a rootstock for its apricots. C and O Nursery is another supplier that carries some apricots on apricot rootstocks. (The addresses for these nurseries appear in the Plant Sources Directory at the end of the book.)

If you are considering buying bare-rooted trees from your local nursery, be sure to inquire about rootstocks. If no one seems to know which ones were used, check the color of the roots for a clue. Apricot roots are redder in color than peach roots, which tend to be brown. This visual inspection may give you something to go on in deciding which tree to purchase.

DO-IT-YOURSELF ROOTSTOCKS

Both of us have decided that perhaps the only way we'll ever get apricots that survive more than a couple of years is to do our own grafting. We're thinking of using seedlings that sprout in our compost pile or that come from the gardens of friends as rootstocks. Besides serving as rootstock materials, these seedlings themselves may one day produce good fruit, for apricots tend to breed true from seed. The only apricot trees Dorothy has at present, other than Manchurian Bush Apricots, are two such seedling trees. Both of them are thriving, and one even produced a few flowers this year after only five years in the garden.

If you live in a problem area for apricots and can't find a source of trees grown on apricot roots, we suggest you learn the basics of grafting and put together your own trees, using apricot seedlings or Manchurian Bush Apricots as your rootstocks. Otherwise, you could just pour money down the drain the way so many other gardeners, including the authors, have done.

APRICOT FLOWERS UP CLOSE

If you examine an apricot tree just before it blooms you'll see that the flower buds are borne on lateral buds of both spurs and one-year-old-wood. Apricot spurs are short lived and usually only blossom for two or three years. The flower buds grow on the sides of the spur and the leaf buds appear at the end, the same arrangement other stone fruits have. The flower buds borne on new wood develop in an interesting pattern. The buds along the base of the shoot produce two or three flowers, while closer to the tip of the stem, the buds form only one flower.

Flower buds for next season are initiated in the late summer months, after the fruit has been harvested. During this period of bud development, apricot trees are especially sensitive to water stress. Even though the fruit has already ripened and you've picked your tree clean, don't ignore it. Apricot trees need plenty of water in late July and August to ensure a bumper crop the following year.

Sometimes, you will see a few flowers develop on your apricot tree after the main flush of blossoms has already appeared and started to form into fruit. This phenomenon occurs if the tree was under water stress in the fall and the following spring turned out

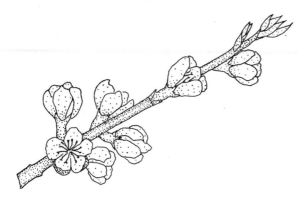

Apricot Blooming: In the spring, look at the way blossoms appear on new wood. Along the base you'll see two or three flowers unfurling from each bud. Closer to the tip, the buds aren't as generous—only a single flower emerges.

to be cool. These errant buds produce a leaf as well as a flower, and the resulting fruits have a long stem and a small pit. The fruit matures slowly, too, and may ripen three weeks after the rest of the crop. This is your tree's way of telling you it didn't get enough water!

HOW TEMPERATURE CAN BE A SPOILER

Apricot flower buds are especially sensitive to changes in temperature, and this sensitivity can lead to much frustration for gardeners in places like California and Oregon. If the weather is warmer than normal in November and December (above 65°F), the slow flower development that normally occurs during dormancy when temperatures are between 32° and 45°F may stop. Since this inhibits the release of the hormones that help keep the buds on the branches, they fall off, robbing you of your harvest.

Temperature fluctuations can take their toll in another way. As winter starts to wane, a spell of unseasonably warm weather can cause apricot buds to deharden. These tender, vulnerable buds are then killed when temperatures sink back down to their more seasonal, cool range. In California, commercial growers sometimes lose their crop when warm weather in December or January is followed by a cold spell. Any radical change in temperature will cause the flower buds to drop off even though the rest of the tree is unaffected.

Since the flower buds deharden so easily, most apricot varieties bloom early in the spring, making them a challenging crop for growers in northern or high-altitude areas which are susceptible to late frosts. Here in western Montana, every five years or so there is a bumper crop of apricots in home gardens; the other four years, few people are lucky enough to harvest any fruit at all since the tender buds have succumbed to a cold snap. To give you some idea of just how touchy apricot flowers are, about 10 percent of the apricot blossoms on a tree will die when the temperature falls to 28°F, but over 90 percent will succumb below 26°F.

IMPORTANT THINGS TO KNOW ABOUT APRICOTS

In most cases, apricot pollination doesn't pose a problem. But growers who live near an aluminum reduction plant may notice that their trees are bearing small crops due to the fact that pollination is being hampered. The culprit is the hydrogen fluoride given off by the industrial plants. This chemical slows down pollen tube growth, which means that fewer flowers are likely to be pollinated.

Like some apples, apricot trees can easily fall into an alternate-year bearing pattern. The only way you can correct this is by thinning very heavily during the period 38 to 41 days after full bloom (when over half of the flowers are open). If you don't thin the fruits at this time, the tree's energy will be sapped by the heavy crop, and it won't be able to produce as many flower buds for the following year. The unfortunate result is that you'll have fewer apricots next season.

If you're a very patient person with a lot of time on your hands, you can go over your trees and thin the apricots so they're 2 to 3 inches apart on the branches. A quicker, easier way to do this thinning was devised by Dr. James Beutel of the University of California at Davis. He suggests taking a length of PVC pipe and *gently* rapping the branches when the fruit is slightly bigger than your thumbnail. You may need to rap several times until the fruit is spaced roughly 2 to 3 inches apart.

Overcoming an alternate-year bearing problem with a tree isn't a once and done affair. You may bring the tree back on an even bearing schedule only to find that the following spring the flowers are killed by frost, which will trigger the tree to produce a heavy set of flower buds for the next year. Then you'll have to thin that year as well. Just keep in mind the delicious reason why you're growing these fruits in the first place!

The amount of sun your apricots receive while they're ripening will determine how much of a rosy glow they have. Fruits exposed to the sun will develop better color than those hidden away in the shade. Sunlight activates a hormone that starts the formation of the red anthocyanin pigments that give the fruit its lovely blush. Be forewarned, though, that there are some varieties that never produce much of a blush, no matter how much sun they get. Moorpark and Sungold are two examples. As with peaches, apricots that have received an excess of nitrogen won't develop much of a red color.

Just as cool temperatures can interfere with flowering, warm weather can have side effects on apricot ripening. If the temperature soars above 95°F while your apricots are ripening, they can develop a problem called pit burn. The heat speeds up the fruit's metabolism so that tissues near the pit can't get rid of their metabolic wastes fast enough and as a result begin to break down. This shows up in susceptible varieties such as Royal as a soft, brownish area right around the pit.

Tilton is less susceptible to pit burn, for it develops an air pocket next to the pit into which the wastes can diffuse. Golden Amber is another resistant variety. Because of the problem of pit burn, many California apricot orchards are located in or near the mountain passes, where the wind that blows through keeps the trees cool.

SOME GENERAL POINTERS ON APRICOT CARE

If you're planting an apricot tree that's grafted onto peach or apricot rootstock, it's critical that you find a well-drained spot. Soggy soil can trigger a self-destructive response by the roots, as we explained earlier under Rootstocks Used for Apricots.

Standard apricot trees don't reach the same towering heights trees like cherries do. You should be able to keep your trees at about 15 feet by cutting the scaffolds back to laterals each winter. What apricots lack in height, they make up for in width, however. Since they can spread quite a bit, it's best to make sure standard trees have 20 feet of clearance in all directions and 24 feet if the soil is especially rich. (It's not always easy to remember this when you're planting a skinny little tree—but they do grow up!) Dwarf apricots need at least 15 feet of growing room.

Prune apricots to an open center as we described under Pruning an Open Center Tree in chapter 6. Because the spurs are so short-lived, you should cut out some older branches each year so that new wood is encouraged to grow. New wood means new spurs to replace the old ones. Always keep the center of the tree open so that the fruit is exposed to sunlight, and cut back any long unbranched branches to a lateral to encourage branching.

WHAT APRICOTS NEED TO GROW

Nitrogen is the most important nutrient for apricot growth. Because the spurs are short-lived and you want your apricot trees to grow new wood each year, you should be sure to guard against any deficiency. One way you can tell whether your trees are getting enough nitrogen is to keep track of shoot growth. The shoots of young trees should grow from 13 to 30 inches each year, while those of bearing trees should grow from 10 to 18 inches.

Apricots need about 0.05 pounds of nitrogen per tree for each year of age starting when they're three years old. Once a tree reaches ten years old, it won't need an annual increase in the amount of fertilizer. (Turn back to The Nutrients Fruiting

Plants Need in chapter 1 for suggestions of good nitrogen sources as well as sources for the other nutrients we discuss here.)

Apricots don't need as much phosphorus as some other fruit trees. If you mulch with compost, your trees should be getting enough. When it comes to potassium, however, apricots often do need an extra boost. The signs to watch for are a lack of vigor, poor bloom and a low set rate.

Remember that apricots need plenty of water, especially as the fruit is ripening and the flower buds for the next year are developing. If you've been lucky enough to find a tree growing on apricot roots, be sure to tend to its water needs. Apricot roots are not very deep, so water regularly and use a mulch to keep the soil evenly moist.

HARVESTING AND STORING APRICOTS

For the best-tasting apricots, you should wait to pluck them from the tree until they're completely ripe. How can you judge when this time has arrived? For starters, the fruit will be a full golden color, with no trace of green remaining. When you gently squeeze an apricot, it should give easily under your finger pressure. Once it's ripe, harvest the fruit by pulling gently upward on it, twisting slightly at the same time so that you don't damage the spurs.

When you're picking apricots for canning, you need to use a slightly different criterion for timing the harvest. These fruits should be firm-ripe so they won't turn to mush when they're canned. If your tree has a bumper crop, you may want to pick apricots for canning first, while they are still a bit firm — not as soft as you would want fruit for fresh eating. Then you could return and pick the slightly riper fruits that are a little bit softer and juicier to use for drying. The best apricots for this purpose have flesh that remains intact when you cut them open. If they're too soft and juicy, they won't dry well. And finally, you could harvest the softest ones for fresh eating, jam making and juicing.

There may come a time when it won't be possible for you to run out to the tree every day and pick the fruits that have just ripened. (Vacations do have a way of coinciding with peak harvesttime!) If you let your apricot tree go unattended, you'll find that as they ripen, the apricots will fall to the ground and won't be fit for anything but the compost pile. Should you plan to be away when your apricots are ripening, your best bet is to pick all your fruit at once. Even though you'll end up with some apricots that are still green, all is not lost. Although the quality won't be quite the same as tree-ripened fruit, you can store and ripen green apricots with some success. Store them under the ideal conditions we discussed in chapter 1 under What to Do with Ripe Fruit until you want them to ripen. Then, keep them at 65° to 75°F until they turn orange and soften. This should happen in a matter of days.

All too often the apricots you buy in the store have a mealy texture and an insipid flavor. While this is partially due to their being picked too early, improper storage conditions are also to blame. Temperatures of 40° to 45°F, which are commonly used for commercial storage, inhibit flavor development and cause the texture to deteriorate.

TIPS ON DRYING AND CANNING THE APRICOT HARVEST

One of the benefits of growing apricots is that you have your own homegrown supply of the golden fruit for drying. Commercially dried apricots are treated with sulfur, and while that helps them retain their bright orange color, naturally dried apricots are just as delicious, minus the questionable chemical treatment.

Drying apricots at home is easy to do. Just cut the fruit in half (or in quarters if it's large), remove the pits, and spread the pieces pit side up on trays outdoors or in a food dryer. If you live in a humid area and want to dry your apricots outdoors, you should

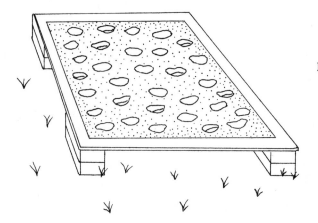

Home-Dried Delicacies: Apricots are easy to dry at home. A spare window screen propped up on stacks of bricks gives air a chance to circulate around all sides of the apricot halves. Be sure to place them pit side up on the screen to dry.

leave them in the direct sun. However, if your climate is hot and dry, you're better off drying the apricots in the sun for only a day or so, then moving them to the shade. That way they'll keep more of their natural color.

Some people let their apricots dry until they reach a leathery consistency. When they're completely dry they can be stored at room temperature. But we like them while they're still somewhat soft, so we store our apricots in the freezer. If we kept them in our cupboards, they'd turn moldy.

Until recently, both of us have been puzzled and discouraged by our experiences with home-canned apricots. In some batches, the texture of most of the fruit is fine, but there are a couple of halves that are too soft. In other batches, the apricots look firm in the jar, but when they're dished out, they almost dissolve into unrecognizable mush.

Now we know the causes for these disappointing experiences. In the case where most of the fruit is fine but some is soft, the culprit is acid. Acid apparently breaks down the cell walls of the fruit during cooking, so the unappetizing apricots in our jars were probably more acid than the others that stayed firm. This sort of softening doesn't increase with storage time.

The other problem—totally dissolving fruit—does get worse with time and is caused by fungi related to the common bread mold. These fungi are commonly found on the surface of apricots and are also associated with the soft brown spots you often see on the skin of ripe fruit. If any of these fungi are present in your batch of apricots, they are destroyed during canning. However, they do produce enzymes that are not totally inactivated by the heat of canning and which can break down the tissues of the fruit. Just a tiny bit of the fungus can turn a whole jar of apricots to mush within nine months. Because of this problem, it's especially important with apricots that you can only completely sound, unblemished fruit. As an extra measure of insurance, you should wash your apricots thoroughly before canning in order to remove any fungi that may be lurking on the surface.

PROBLEMS WITH APRICOTS

Bacterial spot and fungal diseases are the biggest problems for apricot growers, especially in humid regions. Brown rot is one of the most prevalent of these. For pointers on how to deal with brown rot, see Problems with Peaches in chapter 11.

In general, apricots are attacked by the same pests and diseases as peaches but they show more resistance to them. The plum curculio is a major pest of apricots in the East. Another common pest is the gopher; this voracious rodent loves apricot roots, so protect your trees when you plant them as we describe under Getting Fruit Trees into the Ground in chapter 6.

APRICOT FRONTIERS

It's a shame for those of us who want to grow apricots, but researchers aren't concentrating their efforts on developing more and better varieties. Research money tends to go into fruit crops with lots of commercial appeal. Since apples and peaches are more widely grown than apricots, they get the lion's share of the funds available.

This lack of interest in apricots is really too bad, for there are great opportunities for improvement. As we mentioned earlier, most of our varieties were developed from European apricots, which are not as hardy as some Asian types. The Central Asian apricot group, for example, is characterized by vigorous, long-lived trees whose longer winter rest period prevents emergence from dormancy until temperatures warm up in the springtime. In addition, these trees bloom later in the spring than familiar varieties and so are not as likely to be nipped in the bud by a late frost. The Central Asian varieties also have very tasty fruit with sweet, nutlike kernels. One drawback is that they are very susceptible to the fungal diseases brown rot and shot hole, so they can't be grown where the weather is humid.

Guide to Apricot Varieties

Autumn Royal—Medium fruit, late. Zones 5–8. Use fresh, can, dry.

Chinese (Mormon)—Medium, midseason to late. Zones (4)5–8. Good for late frost areas. Needs pollinator. Use fresh, dry.

Early Golden—Medium to large, early to midseason. Zones 5–9. Use fresh, can.

Garden Annie—Medium to large, early. Zones 5–8. Genetic dwarf, 5–8 feet tall.

Goldcot—Medium to large, late. Zones 5–8. Use fresh, can, dry.

Golden Amber—Large, late. Zones 5–8. Resistant to pit burn. Use fresh, can, dry.

Golden Glo—Late. Zones 5–8. Genetic dwarf, 4–6 feet tall. Use fresh, dry.

Gold Rich—Large, midseason. Zones 5–8. Needs pollinator.

Manchurian Bush Apricot—Small, juicy, sweet. Zones 4–9. Bush type.

Moongold—Medium, midseason. Zones (4)5–8. Dwarf, needs Sungold as pollinator.

Another potential source of useful breeding material is the Irano-Caucasian group of apricot trees from the eastern Mediterranean area. These trees tend to have very sweet white or light yellow fruit with edible kernels. They are adapted to mild winters and begin to grow fairly early in the spring, traits that certainly would appeal to southern growers. There are also other groups of apricots which could profitably be crossed with our varieties to improve them—if only the interest could be generated among researchers for the necessary breeding to be done!

Moorpark—Extra large, midseason. Zones 5–9. Good pollinator. Use fresh, can, dry.

Perfection—Large. Zones 6–8. Low chill, needs pollinator. Use fresh, can.

Riland—Medium, midseason. Zones 6–8. Needs pollinator.

Rival—Extra large, early to midseason. Zones 6–8. Early blooming. Needs pollinator. Can.

Royal (Blenheim)—Large, early to midseason. Zones 5–8. Use fresh, can, dry.

Scout—Medium to large, midseason. Zones (4)5–8. Good for northern growers.

Sungold—Medium, midseason. Zones (4)5–8. Dwarf, needs Moongold as pollinator. Use fresh.

Tilton—Medium, midseason. Zones 6–8. High chill, good pollinator. Can.

PLANT SOURCES DIRECTORY

Ahrens Strawberry Nursery
Huntingburg, IN 47542
  Concentrates on strawberries but also
  carries brambles, some grapes, blueberries and fruit trees. Catalog contains
  some helpful information about varieties.

Bountiful Ridge Nurseries
Princess Anne, MD 21853
  Specializes in fruits and nuts; offers many
  small-size trees and often a choice in
  rootstocks.

Brookdale Nursery
Box 422
Newcastle, New Brunswick,
Canada E1B 3M5
  Carries lowbush blueberries.

C and O Nursery
Box 116
Wenatchee, WA 98801
  Supplies mainly fruit trees, with a
  few small fruits and nuts. Catalog
  gives information about rootstocks,
  pollination and harvesting dates.

Carlton Nursery
Rt. 1, Box 224
Dayton, OR 97114
  Carries cling peaches, pears on OHxF
  rootstocks and oriental pears. Mainly a
  commercial supplier, but will fill small
  orders to individuals.

Cumberland Valley Nurseries
Box 430
McMinnville, TN 37110
  Specializes in peaches, nectarines and
  plums.

Dean Foster Nurseries
Hartford, MI 49057
  Carries an assortment of fruits, with
  emphasis on strawberries. Catalog has
  helpful information on varieties.

Emlong Nurseries
Stevensville, MI 49127
  Offers hardy varieties for the North.

Foster Nursery
Fredonia, NY 14063
  Carries a wide selection of grapes, with
  a handy and informative "Guide to
  American and French Hybrid Grape
  Varieties" available for $1.50.

Hilltop Orchards and Nurseries
Rt. 2
Hartford, MI 49057
  Mainly a commercial supplier, but will
  fill home orders for five or more trees.
  Uses Manchurian rootstock for apricots.
  Catalog includes information on harvest-
  ing, rootstocks and pollination, with
  helpful descriptions of varieties.

Ison's Nursery and Vineyard
Brooks, GA 30205
  Supplies fruits and nuts for the South,
  including many muscadine grape
  varieties.

Lawson's Nursery
Rt. 1, Box 294
Ball Ground, GA 30104
  Specializes in unusual and antique
  varieties, especially of apples.

Henry Leuthardt Nurseries
East Moriches, NY 11940
  Provides helpful guidebook on dwarf
  fruit trees.

J. E. Miller Nurseries
Canandaigua, NY 14424
  Offers wide variety of fruits, especially
  grapes and apples, including many
  antique apple varieties. Specializes
  in fruits for the North.

New Brunswick Horticultural Nursery
Hoyt, New Brunswick, Canada E0G 2B0
  Carries lowbush blueberries.

North American Fruit Explorers
Ms. Mary Kurle
10 South 055 Madison St.
Hinsdale, IL 60521
  Discusses new development in fruit
  plants, including new varieties and

advances in breeding. The journal, *Pomona,* presents information of interest to backyard fruit growers. Dues $6.00 ($8.00 in Canada).

Owen's Vineyard and Nursery
Georgia Highway 85
Gay, GA 30218
    Carries varieties for the South, especially muscadine grapes. Catalog contains information on growing rabbiteye blueberries.

Patrick's Vineyard, Orchard Nursery and Farm Market
Pomegranate Boulevard
TyTy, GA 31795
    Supplies fruits and nuts for the South, including many exotic fruits.

Peaceful Valley Farm Supply
11173 Peaceful Valley Rd.
Nevada City, CA 95959
    Offers genetic dwarfs as well as grafted dwarfs. Also carries five varieties of Asian pears, good selection of western trailing blackberries, and fruit trees that

are grown organically. Catalog gives information on rootstocks.

Rayner Brothers
Box 1617
Salisbury, MD 21801
    Specializes in strawberries, but also carries some other fruits.

Southmeadow Fruit Gardens
Lakeside, MI 49226
    Offers a tremendous number of varieties, many of which are hard to find. Carries lots of antique apple varieties and many more pear varieties than any other nursery.

Stark Brothers Nurseries
Louisiana, MO 63353
    Carries a wide variety of fruits for all sections of the country.

Maudie Walker
Box 256
Omega, GA 31775
    Carries fruits for the South, especially muscadine grapes and rabbiteye blueberries.

GLOSSARY

Abscission layer: a special layer of cells that forms on leaf petioles and fruit stems; this layer cuts them off from the tree so that leaves and ripefruit will fall.

Anther: the top part of the stamen that contains the pollen.

Apical tip: the region at the tips of branches and spurs where new growth occurs; this area also produces hormones that control growth in other parts of the plant.

Auxin: a plant growth hormone produced by young leaves, seeds in fruit and apical tips; this hormone is responsible for such phenomena as the branching pattern of trees and fruit growth.

Axillary bud: a bud located in the leaf axil (the angle between the leaf and the stem) which serves as a potential growth site.

Chill requirement: the number of hours at temperatures within a critical range, generally between 32° and 45°F; each plant and each particular variety has a different chill requirement; plants with high chill requirements need more hours of cold than do those with low chill requirements.

Clone: offspring which is genetically identical to the parent plant; obtained through vegetative reproduction, which means that each new plant is formed from tissues of the parent plant.

Crown: the area where the stem(s) and roots meet.

Dormancy: the period in a plant's annual growth cycle that lasts from the time it loses its leaves in the fall until the flower or leaf buds begin to open in the spring.

Drip-line: the imaginary circular area on the ground beneath a tree that corresponds to the reach of the longest branches.

Genetic dwarf: a tree that naturally grows to a small size because of its genetic makeup, without any need for grafting onto a dwarfing rootstock.

Grafting: a method used to join parts of plants together so that they will unite and grow as one plant.

Hybrid: a plant produced by crossing two different species.

Node: a region of the stem from which leaves and axillary buds arise.

Phloem: the conducting tissue in a plant stem that carries hormones and food from the leaves throughout the rest of the plant.

Pistil: the female organ of a flower which consists of a stigma, style and ovary.

Pollenizer variety: a compatible fruit tree capable of pollinating neighboring trees of a different variety; also called pollinator variety.

Pollinator: an agent such as the wind or a bee that transfers pollen from one flower to another; also see pollenizer variety.

Rest period: the period in a plant's annual growth cycle during which it must receive cold temperatures; new growth will not occur until this low temperature requirement has been met; the period lasts from the time a plant loses its leaves until it has awakened and is ready to blossom or leaf out.

Rhizome: a horizontal underground stem that can produce upright shoots with the ability to grow through the soil and develop into new plants.

Rootstock: the lower part of a grafted plant which provides the root system for the plant.

Self-fruitful: a plant that is able to pollinate itself (or be pollinated by another plant of the same variety) and bear fruit.

Self-unfruitful: a plant which can only produce fruit by cross-pollinating with another variety.

Set rate: the number of fruits that start to grow on a plant after the flowers have bloomed.

Spur: a short, compact branch that has the potential to bear both fruit and leaves; the place where fruit is borne on many fruit trees.

Stamen: the male organ of a flower consisting of an anther and filament.

Sucker: extra growth in the form of a shoot that appears on a plant; generated by buds below the ground, usually on roots.

Xylem: the conducting tissue in a plant stem that carries water and minerals from the roots to the rest of the plant; it also transports hormones.

INDEX

Page references in italics indicate illustrations.

DATE DUE
